ACMTAA-2013

Advanced Composite Materials and Technologies for Aerospace Applications

Richard Day • Sergey Reznik • Thuc Vo • Tahir Sharif • Yuriy Vagapov

ACMTAA-2013

Proceedings of the Third International Conference on

Advanced Composite Materials and Technologies for Aerospace Applications

May 13-16, 2013, Wrexham, North Wales, United Kingdom

Edited by

Prof Richard Day
Glyndŵr University

Prof Sergey Reznik
Bauman Moscow State Technical University

Dr Thuc Vo
Glyndŵr University

Dr Tahir Sharif
Glyndŵr University

Dr Yuriy Vagapov
Glyndŵr University

2013 • Wrexham • Glyndŵr University

Proceedings of the Third International Conference on
Advanced Composite Materials and Technologies for Aerospace Applications
May 13-16, 2013, Wrexham, North Wales, United Kingdom

Edited by

Prof Richard Day
Glyndŵr University

Prof Sergey Reznik
Bauman Moscow State
Technical University

Dr Thuc Vo
Glyndŵr University

Dr Tahir Sharif
Glyndŵr University

Dr Yuriy Vagapov
Glyndŵr University

Published in 2013 by
Glyndŵr University
Plas Coch
Mold Road
Wrexham
LL11 2AW
United Kingdom
www.glyndwr.ac.uk

ISBN 978-0-946881-80-2

Printed and bound in the United Kingdom

ACMTAA-2013

The Third International Conference on Advanced Composite Materials and Technologies for Aerospace Applications

May 13-16, 2013, Wrexham, North Wales, United Kingdom

The conference is organised by

Bauman Moscow State Technical University, Russia
Glyndŵr University, United Kingdom

The conference is supported by

Institute of Physics (IOP) in Wales

The conference co-chairmen

Prof Richard Day
Glyndŵr University, United Kingdom

Prof Sergey Reznik
Bauman Moscow State Technical University, Russia

International Programme Committee

Prof Alison McMillan
Glyndŵr University
United Kingdom

Prof Antonio J. M. Ferreira
University of Porto
Portugal

Dr Ali Hasan Mahmood
NED University of Engineering and Technology
Pakistan

Dr Bilal Zahid
NED University of Engineering and Technology
Pakistan

Prof Costantinos Soutis
University of Manchester
United Kingdom

Dr Fawad Inam
Glyndŵr University
United Kingdom

Prof Galina Malysheva
Bauman Moscow State Technical University
Russia

Prof Jaehong Lee
Sejong University
Korea

Dr Lukasz Figiel
University of Limerick
Ireland

Dr Muhammad Dawood Hussain
NED University of Engineering and Technology
Pakistan

Prof Xiao (Matthew) Hu
Nanyang Technological University
Singapore

Dr Zhongwei Guan
University of Liverpool
United Kingdom

Local Organising Committee

Dr Tahir Sharif (Chair)
Dr Thuc Vo (Proceedings, Paper Submission and Review Manager)
Dr Yuriy Vagapov (Proceedings Manager)
Mr Cedric Belloc (Industrial and Professional Bodies Liaison Manager)
Mrs Olga Edwards (Visa Support, Logistics and Accommodation Manager)

Content

Content

Advanced Directions of Research in the Field of Composite Structures for Space Antennas

Sergey Reznik

Rocket and Spacecraft Composite Structures Department, Faculty of Special Machinery, Bauman Moscow State Technical University, 5 2nd Baumanskaya Street, Moscow, 105005, Russia

Abstract: Fabrication of space reflectors is a complex interdisciplinary problem, covering the fields of radio engineering, materials science, thermal and stress analysis. A number of tendencies justified the demand for innovative engineering solutions, among them 15 and more years' increase in the satellite life cycle, higher frequency radio signal, greater information capacity of the payload and larger reflectors. The issue of designing composite structures for space reflectors with rigid and mesh surfaces is investigated. Mathematical modeling data, experimental methods and results are presented.

Key Words: Space antennas, Composite materials and structures, Metal meshes, Mathematical modelling, Tests.

1. Introduction

Composite materials (CM) have found increasingly wide application in aerospace industry since the middle 1950s owing to their high strength to weight ratio, stiffness, long -term resistance to extreme temperatures, to high pressure, to reactive and erosive flow, to particulate radiation and other external factors.

One of the characteristic features of CM is that they cannot be viewed separately from the structure and the manufacturing technology. To date there have been developed complex parts fabrication methods from CM with polymer, metallic, carbon, or ceramic matrices reinforced with particles, fibres or fabrics from organic, non-organic or metallic materials. CM have been successfully utilized for the heat shield in manned and unmanned space vehicles (SV), for the solid rocket motors, for high-pressure vessels, rocket engines nozzles, space antennas, solar cells and energy systems, payload fairings, wing leading edges and trim panels.

However, most types of CM continue to be more expensive than the traditional metals and alloys, composite technology is characterized by low energy efficiency, and the equipment is extremely complex and material intensive. The development of new technologies and there transfer from laboratory to manufacture is still largely done by means of intuitive-empirical methods.

Nevertheless, the recent trends indicate the emergence of new powerful research tools capable of eliminating these shortcomings. First of all, it became possible due to the development of new high performance software to model complex manufacturing procedures without simplifying assumptions, i.e. in 3D including the dynamics of the processes in the materials and reaction space. Methods were developed to modify CM by embedding nanoscale particle into the matrix or filler, which resulted in superior physical properties of the CM. A number of existing procedures, e.g. infusion, were refined to reduce material intensity and increase the energy efficiency of the polymer CM structures.

2. Means and methods of theoretical and experimental research

Design of composite structures for rocket space applications is based upon theoretical analysis of thermal and stress-strain behaviour. Apart from standard ANSYS, NASTRAN, PATRAN, SINDA and SINARAD software, Bauman MSTU researchers use an original package of CAR software to solve direct and inverse problems of combined (predominantly radiative-conductive) heat transfer (Reznik, 1992; 1996; 2001; 2003). The range of problems covered by this package include:

- predicting temperature regimes for thermally stable structures from opaque and semitransparent materials under space flight conditions both for atmospheric and ground tests;
- planning temperature measurement for ground and flight tests; analysis of methodical errors of temperature measurement with sensors;
- identifying heat transfer parameters (thermal physical, volumetric and surface optical characteristics) for isotropic and anisotropic materials by experimental data obtained through coupon and full-scale elements testing on various testing setups and facilities.

Ground testing is in essential stage in the space antennas manufacturing process. It includes laboratory testing of materials and coatings samples, determining their properties and resistance to the outer space factors. Bench testing involves mock-ups and full-scale structures under simulated flight conditions, design and systems performance are tried and tested. Bauman MSTU in collaboration with the partners designed a number of new testing methods (Reznik, 2002). Their characteristic features include:

- computer-aided tests preparation and conduction and data processing ;
- uniform and spatially confined heating to form non-uniform thermal fields in the test objects;
- using noncontact and contact temperature measurement devices.

There is insufficient reference data on thermal physical and optical characteristics of CM, with the total lack of data on metal meshes characteristics, which resulted in designing innovative testing methods (Reznik and Denisov, 2008; Reznik et al, 2010; Denisova et al., 2011; Reznik et al., 2011).

The method of determining coefficient of thermal conductivity for CM involved one-sided heating of full-scale rod structure elements in a vacuum chamber, measuring longitudinal temperature distribution and processing the testing data by means of software for the nonlinear inverse problem of thermal conductivity. The obtained temperature dependences for thermal conductivity of carbon fibre reinforced plastics (CFRP) cover a broad range of temperatures and radically differ from the data available before research.

Spectral optical characteristics of metal meshes, such as transmitivity A_v, reflectivity R_v and emissivity ε, with v for frequency, were determined by means of standard optical devices, while integral parameters, such as A/ε, were determined through thermal tests in vacuum chamber with solar simulation (Reznik et al, 2010 Reznik et al., 2011). The temperature dependence for the specific heat capacity of metal mesh was determined with standard devices, such as IT-c-400. The implementation of these methods increased the accuracy of thermal calculations.

3. Research directions and outcomes

In order to solve new challenges in the fields of space navigation, geology, exploration of planets, their satellites and other celestial bodies it is essential to further improve space antennas. It mainly concerns reflector antennas, both rigid reflectors and deployable reflectors of the lobe, umbrella or stretched membrane types. The efficiency of the antenna (the number of beams and its pointing accuracy) grows with the area of reflector; however, it stipulates the requirements to the precision of the reflecting surface profile. It is known that the acceptable deviation of the profile should not exceed the magnitude of $\Lambda/16$, where Λ is the antenna's operating wavelength (Gryanik and Loman, 1982).

One of the targets in reflectors design is to achieve minimum mass per unit length (ratio the reflector mass to the aperture area). According to *ESA*, this parameter is markedly different for various reflector types (Fig. 1) (Prowald et al, 2011).

A number of communication satellites, such as Yamal-300K, Yamal-401, Enisei-A1 (previously Luch-4), Amos-4 (Israel), Libid (Ukraine), which were developed in *ISS-Reshetnev*, as well as Inmarsat, Intelsat satellites by *Boeing*, Astra satellites by *EADS Astrium* and others employ rigid reflectors operating the frequency band of C (4-8 GHz, Λ=75-37.5 mm); Ku (12-18 GHz, Λ=25-16.7 mm); Ka (27-40 GHz, Λ=11.1-7.5 mm). For instance, Inmarsat-5 satellite is equipped with two transmitting and four receiving antennas with 89 beams in Ka-band. It is evident that the acceptable deviation in the reflector profile constitutes fractions of millimeter.

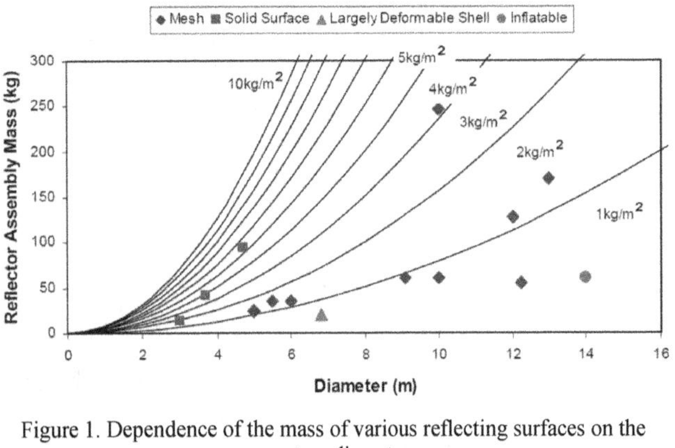

Figure 1. Dependence of the mass of various reflecting surfaces on the aperture diameter.

Since the new generation satellites are meant to operate in the orbit for at least 15 years, there appears a challenge of preserving the shape and the dimensions of the reflector under the outer space conditions and periodic changes of temperatures caused by entering the Earth's shadow zone. The best materials for such reflectors are CFRP, with the low values of linear thermal expansion, high specific strength and stiffness.

Fig. 2 and Fig. 3 display the results of mathematical modeling of thermal and stress-strain behavior of a solid surface antenna reflector, which was performed by means of PATRAN, SINDA and SINDARAD software by *MSC.Software* and *Siemens* FEMAP and NX NASTRAN software.

It was assumed that a reflector is in the orbit and it is exposed by a solar radiation flux with 1400 W/m^2 density at 45° to the parabolic shell surface. The aperture of the

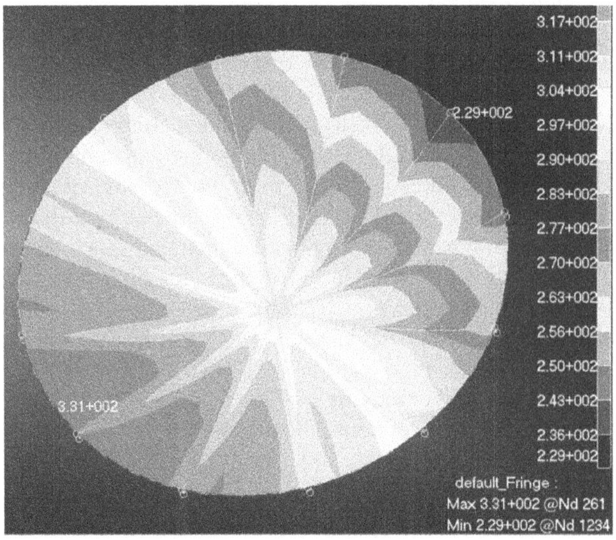

Figure 2. Thermal state of a rigid reflector of a space antenna (temperature is given in K).

Figure 3. Stress-strain behaviour of a rigid reflector of space antenna (displacement is given in mm).

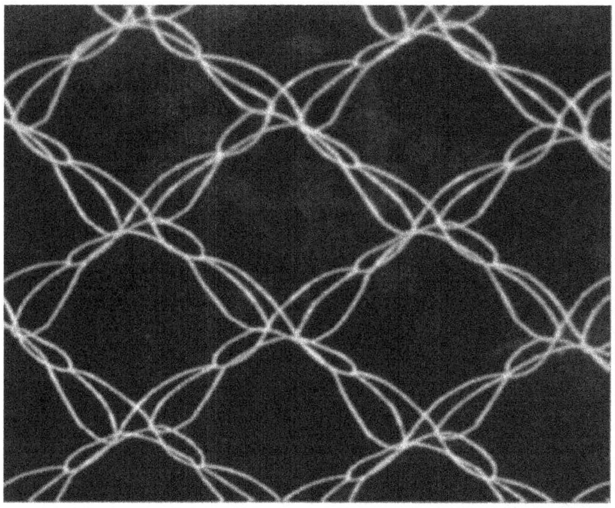

Figure 4. Metal mesh fabric structure.

reflector shell constituted 1 m; the wall thickness was 1 mm, with 2 mm thickness of each of the supporting ribs. It was fabricated from four monolayers of carbon reinforced plastic with the 0/+45/–45/90 degrees layup angles.

As evident from the diagram the temperature change between the illuminated and shaded area of the reflector surface constitute more than 100 K and can cause thermal stresses and strains. The stiffness analysis indicated that the dimensional stability of this structure is inferior to that of an isogrid or a radial ring structure, with less higher natural frequencies. The results of calculations are used as a basis for the selection of the layout.

Russia, the US and Europe are working on developing large deployable mesh reflectors. In particular, umbrella type mesh reflectors are installed on board geostationary spacecrafts-retransmitters of Luch-5 series by *ISS-Reshetnev* and on board TDRS, ACeS, SDARS satellites by *Harris* (USA) (Geostationary spacecrafts-retransmitters, 2008; Thompson, 2002;). The largest stretched membrane reflectors were developed by *AstroMesh* (USA) for Thuraya-3, MBCO, Inmarsat, GlobalStar (Marks, 2011). Luch-5B satellite is equipped with two reflectors 4.2 m in diameter which can be directed at low-orbit space antennas, to capture and guide them along the flight path. One of these reflectors operates in Ku-band and the other in S-band.

According to *European Space Agency* (Prowald et al, 2011), the area density for mesh reflectors is in the region of 1 kg/m^2, which signifies high mass-geometry efficiency. These high values could be achieved owing to thin-wall rod and plate elements from CFRP in the backing structure combined with metal meshes woven from mo-

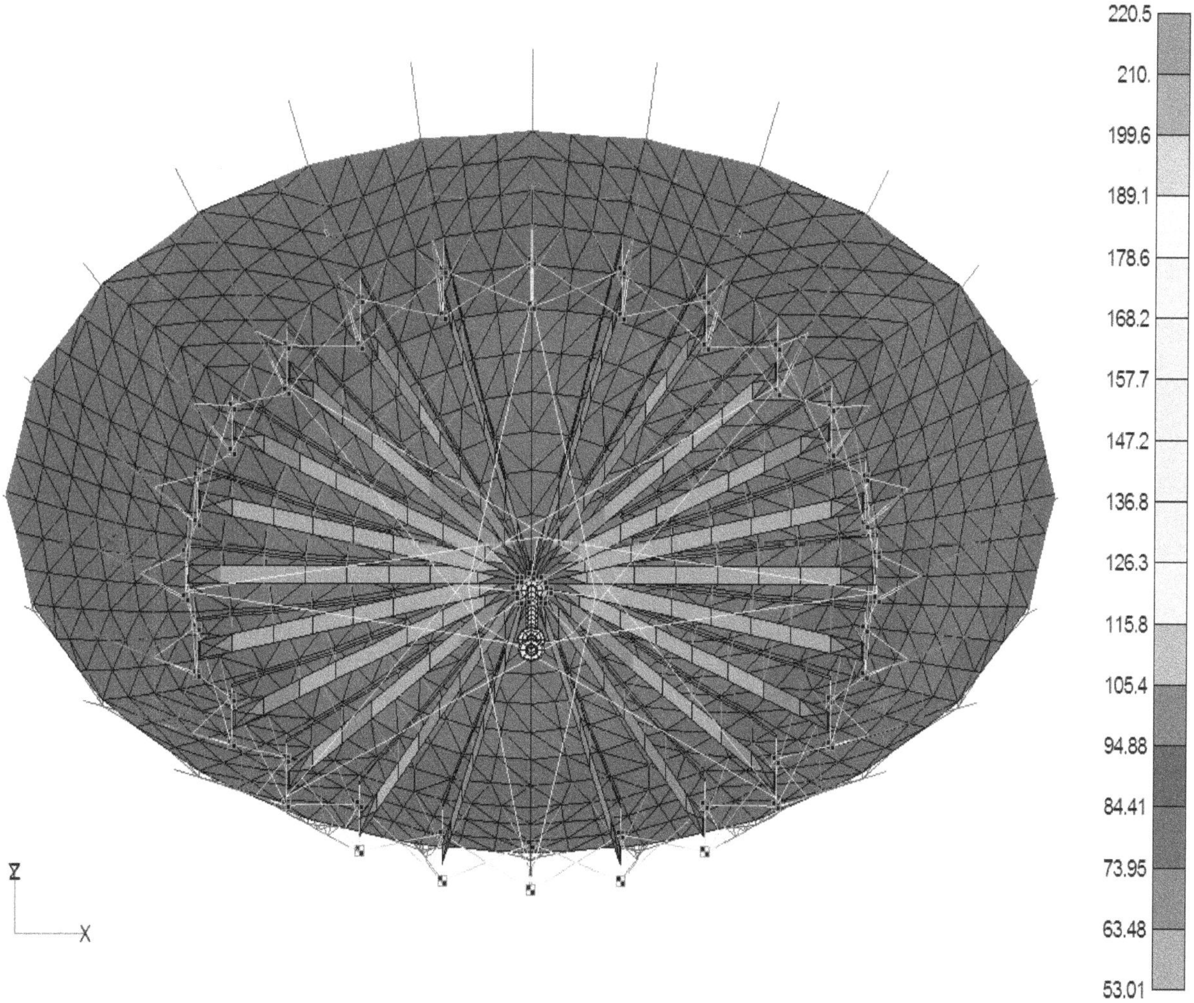

Figure 5. Temperature distribution in a deployed reflector. Spring equinox. The end of the shadow zone.

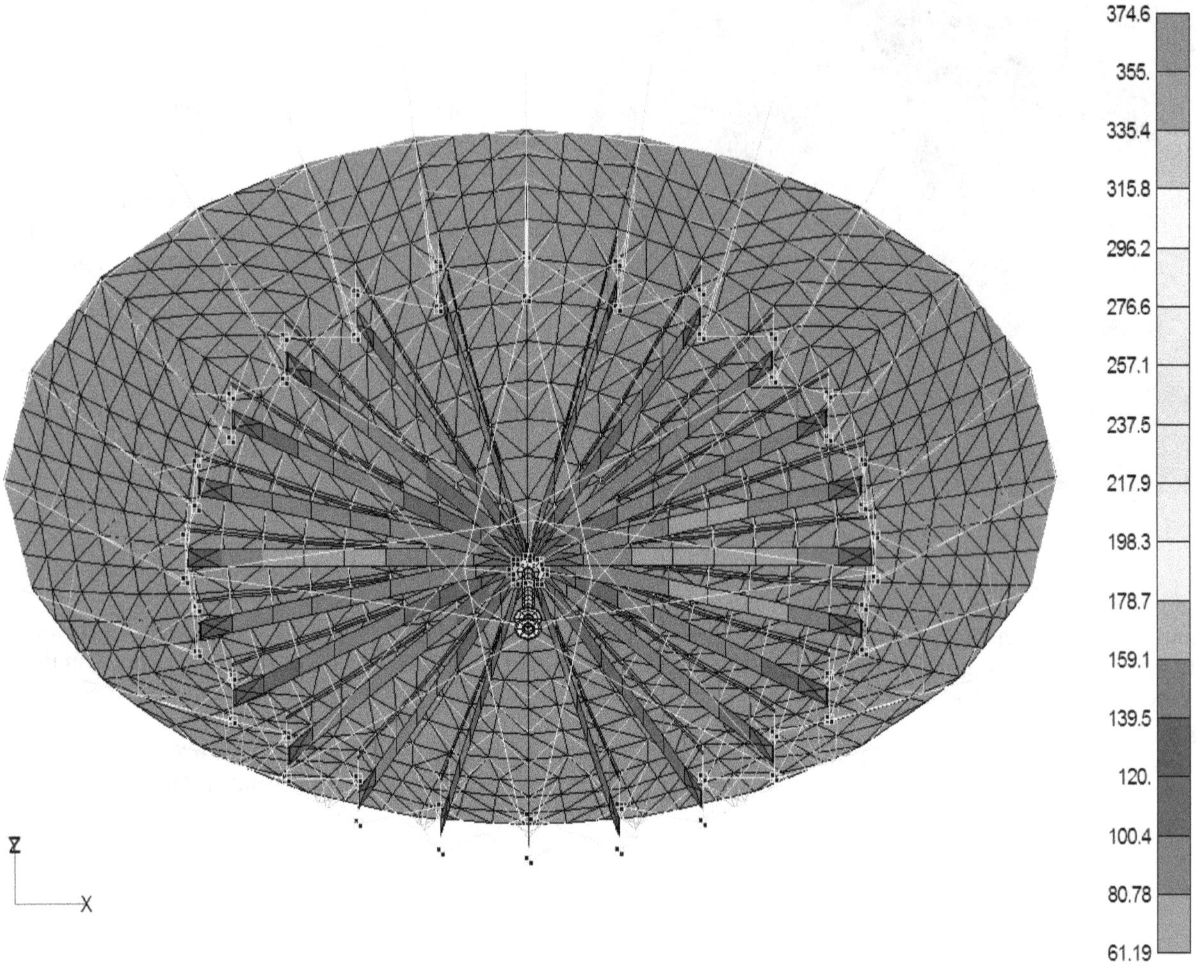

Figure 6. Temperature distribution in a deployed reflector. Winter solstice. Sun capture.

lybdenum, tungsten or nichrome wire 15-50 μm in diameter, gold-plated to improve radio reflecting properties (Fig. 4) and organic fiber cables (Zavaruev et al., 2007).

The joint experience of the *Energiya Space Corporation* and *Alenia Spazio* (now *Thales Alenia Space*) bears evidence to the challenges in the design of mesh reflectors for large antennas (Gottero et al., 2005). The specific feature of the structure is a circular deployable truss formed by hollow CFRP rods in combination with flat ribs from net-shaped CFRP. The reflecting surface is manufactured from three layer mesh fabric from gold-plated tungsten wire 15 μm in diameter.

As regards the thermal calculations, mathematical modeling of such reflectors is challenging due to the complexity of the structure, various materials used with their optical and thermal physical properties not fully investigated. The calculations are time consuming because the radiant heat transfer among several thousands of surface elements, which constitute the geometrical model, must be accounted for (Reznik et al., 2008, 2010).

The characteristic feature of the geostationary orbit is the period when the space antenna is in the Earth's shadow, which lasts up to 71 minutes at the spring and autumn equinoxes. Consequently, the thin-wall structure elements (mesh, ribs, cables) can cool to very low temperatures by the end of the shadow period which was not at the least anticipated by the designers (Fig.5, Fig.6). However, the calculations of the reflector's structure for the shadow part of the geostationary orbit which were performed by means of the finite-elements CAR software were verified by the calculations done in *TSNIIMASH* and *Alenia Spazio* and by experiments in *ESTEC* (Gottero et al., 2005). It is clear in terms of physics that thin-wall elements cannot accumulate a large amount of heat over the illuminated part of the orbit and on entering the shadow zone the temperature of the exposed elements is to decrease approaching the 4 K temperature of the outer space.

Numerical simulation of thermal regimes for reflectors identified a number of factors to be taken into consideration:

- non-uniform exposure of the parabolic reflectors to solar radiation fluxes causes periodical self-shadowing of the inner surface areas;
- metal mesh reflector shells can be viewed as "thermally thin bodies" from semi-transparent material;
- the application of decomposition principle with separate analysis of thermal regimes for reflector and SA depends on the specific structural layout. Re-radiation from the SA frame accounts for the increase of the reflector temperature into the results of the calculations;

- disregard of the terrestrial thermal radiation at the shadow areas of the geostationary orbit results in significantly (by 40-50 K) lower temperature values of the thin elements, like thin ribs with a small area of the irradiated surface and a large area deflecting the heat outside.

Conclusion

1. Critical goals in the fields of communication, navigation, Earth remote sensing and astrophysical research presuppose longer operating life cycle of spacecraft equipped reflector antennas, which should have superior dimension stability in the traditional and prospective radio frequency bands compared to the existing structures.
2. The experience to date suggests feasibility of utilizing polymer composites materials for space antennas with rigid, mesh or flexible adaptive surface of reflectors.
3. Available CAD systems and computer simulation software for space antennas composite structures require improvement both in terms of investigating the material properties CM and metal meshes in the full range of operating environments, and in full-scale structures simulation tests.

Acknowledgments

A number of results presented in the paper were obtained due to financial support of the RFBR grant No. 12-08-00305-a. The author is grateful to his colleagues – Andrei Azarov, Liliana Denisova, Oleg Denisov, Pavel Prosuntsov, and Inna Shafikova for their scientific cooperation.

References

Denisova, L.V., Kalinin, L.V. and Reznik, S.V. (2011). Theoretical and experimental investigations of thermal regimes for mesh reflectors of space antennas, Herald of Bauman MSTU, Machine Building, No. 1 (82), pp. 92-105. (in Russian).

Gottero, M., Sacchi, E., Scialino, G.L., Reznik, S.V. and Kalinin, D.Yu. (2005). The large deployable antenna (LDA). A review of thermal aspects. *Proceed. 35th Int. Conference on Environmental Systems,* (Rome, Italy, 11-14 July 2005), No. 05ICES-369.

Gryanik, M.V. and Loman, V.I. (1987). *Deployable Umbrella Type Antennas.* Moscow: Radio i Svyaz Publishing House. (in Russian).

Geostationary spacecrafts-retransmitters "Louch-5A" and "Louch-5B". (2008). http://www.iss-reshetnev.ru/?cid=prj_ca_loutch5a. (in Russian).

Marks, G., Lillie, C. and Kuehn, S. (2011). Application of the AstroMesh reflector to astrophysics missions (Zooming in on black holes). *Proceed. 33rd ESA Antenna Workshop on Challenges for Space Antenna Systems. Preparing for the Future*. (Noordwijk, TheNetherlands, 18-21 October, 2011).

Prowald, J., Mangenot, C., Klooster, van't K., Scolamiero, L. (2011). Large reflector antennas: technical and programmatic status one year after the working group conclusions. *Proceed. 33rd ESA Antenna Workshop on Challenges for Space Antenna Systems* (Noordwijk, The Netherlands, 18-21 October, 2011).

Reznik, S.V. & Kalinin, D.Yu. (2003). *Modeling of the Thermal Regimes of Large Space Structures.* Bauman MSTU Publishing, Moscow, 52 p. (in Russian).

Reznik, S.V. (1992). Radiative and conductive heat transfer in large-size space structures. *Proceed. 2nd Minsk Int. Forum on Heat and Mass Transfer – MIF-92,* (Minsk, Belarus, 18-24 May, 1992). Vol. 2, pp. 195-198. (in Russian).

Reznik, S.V. (1996). Modern Problems of Modeling and Identification of Radiative and Conductive Heat Transfer. *Proceed. 3rd Minsk Int. Forum on Heat and Mass Transfer – MIF-96,* (Minsk, Belarus, 20-24 May, 1996). Vol. 2, pp. 141-149. (in Russian).

Reznik, S.V. (2001). Modeling and inverse problems of radiative and conductive heat transfer. *Proceed. Eurotherm Seminar 68 on Inverse Problems and Experimental Design in Thermal and Mechanical Engineering.* (Poitiers, France, 5-7 March, 2001), Futuroscope, Poitiers, pp. 23-35.

Reznik, S.V. (ed.) (2002) *Materials and Coatings for Extreme Performances. Prospection.* In 3 volumes. Moscow: Bauman MSTU Publishing. (in Russian).

Reznik, S.V. and Denisov, O.V. (2008). Setup and results of thermal tests of elements of composite rod space structures, *Herald of Bauman MSTU,* No. 4, pp. 81-89. (in Russian)

Reznik, S.V., Kalinin, D.Yu. and Denisov, O.V. (2008). Features of large deployable antennas thermal state in space. *Proceed. 30th ESA Antenna Workshop on Antennas for Earth Observation, Science, Telecommunication and Navigation Space Missions* (27 - 30 May 2008, ESA/ESTEC Noordwijk, The Netherlands).

Reznik, S.V., Kalinin, D.Yu. and Denisova, L.V. (2010). Modeling of metal meshes thermal regimes for space antennas. *Proceed. 32th ESA Antenna Workshop on Antennas for Earth Observation, Science, Telecommunication and Navigation Space Missions.* (Noordwijk, The Netherlands, 5-8 October 2010).

Reznik, S.V., Denisova, L.V. and Mikhailovskiy K.V. (2011). Thermal tests of metallic mesh elements for space antenna reflectors. *Proceed. 33rd ESA Antenna Workshop on Challenges for Space Antenna Systems. Preparing for the Future*. (Noordwijk, TheNetherlands, 18-21 October, 2011).

Thomson, M.W. (2002). The AstroMesh deployable reflectors for KU- and KA- band commercial satellites. *AIAA Papers.* No. 2032.

Zavaruev, V.A., Kudryavin, L.A. and Khalimanovich, V.I. (2007). Knitted metal mesh fabric for the reflecting surface of deployable ground and space antennas. *Technical Textiles.* No.16, pp. 59-64. (in Russian).

Rapid Microwave Processing of Epoxy Nanocomposites Using Carbon Nanotubes

Nataliia Luhyna, Fawad Inam, Ian Winnington

Advanced Composite Training and Development Centre and Department of Engineering, Glyndŵr University, Plas Coch, Mold Road, Wrexham, LL11 2AW, United Kingdom

Abstract: Microwave processing is one of the rapid processing techniques for manufacturing nanocomposites. Though there is considerable amount of work done on the development of carbon nanotube (CNT) based nanocomposites, there is very little work focussing on the addition of CNTs for shortening the curing time. Using microwave energy, the effect of CNT addition on the curing of epoxy nanocomposites was researched in this work. Differential scanning calorimetry (DSC) was used to determine the degree of cure for epoxy and nanocomposite samples. CNT addition significantly reduced the duration for complete curing of epoxy nanocomposites. As compared to monolithic cured epoxy, 20.5% of decrease in time and 12.5% decrease in spent consumed energy were observed for 0.2 wt.% CNT filled epoxy nanocomposite.

Key Words: Carbon nanotubes, Epoxy nanocomposites, Microwave curing.

1. Introduction

For manufacturing of advanced composites, researchers and industrialists are always seeking ways to reduce product cost and environmental damage. In this regard, energy and time spent for the manufacturing of a product are the most significant parameters. Therefore, it is important to research innovative technologies available for rapid curing of composites. Some of the attractive options as compared to conventional curing are UV methods, gamma radiation, microwaves and electron beam. The most productive of all are microwave and electron beam processing, both based on electromagnetic oscillations. Electromagnetic oscillations can be divided into two main groups: high frequency currents and ultra-high frequency or microwave radiation. Electron beam has many advantages (Sui et al., 2000; Wolff-Fabris, 2010) but for curing composites it is responsible for imparting large shrinkage and low glass transition temperature to the thermoset composites (Ghosh and Palmese, 2005). Microwave radiation is electromagnetic radiation with a wavelength in the range of 1 mm (300 GHz) to 30 cm (1 MHz) (Pinprayoon, 2007).

Microwave energy is a very attractive option because it uses less energy and less time to produce composites (Meyer and Herbeck, 2005; Das et al., 2009; Ku and Yusaf, 2008; Yusoff et al., 2007). The application of high -frequency electric fields for curing the thermosetting composites is a rapid processing technique and it is well known that microwave treatment has numerous advantages compared to thermal heating (Wallace et al., 2006; Rangari et al., 2010; Chang et al., 2012). The technique offers deep and uniform penetration of microwaves into the sample, capability of preferential heating sites and rapid production rate (Judith, 1999; Zhou and Hawley, 2003). It is worth mentioning here that microwave heating occurs thoroughly through the entire thickness of the sample. Uniform curing through the thickness of the composite is quite complicated or impossible to achieve during conventional thermal heating (Judith, 1999; Pinprayoon, 2007). Due to its simplicity and effectiveness, microwaves can also be used to restore any damage in a composite structure (Zhang and Dai, 2006). Microwaves enable to achieve the finished composites in a few minutes maintaining the mechanical properties, and sometimes even exceed it as compared to several hours long conventional thermal curing (Chang et al., 2012; Papargyris et. al, 2008). Zhou and Hawley (2003) reported stronger crosslinked bonds in thermoset polymers cured by microwaves. The glass transition temperature is much higher for microwave cured samples as compared to those cured with the conventional heat (Wallace et al., 2006).

Looking into the kinetics of microwave curing, the physical cause of the existence of the electromagnetic field is due to the time-varying electric field which creates magnetic field and the changing magnetic field generates the vortex electric field (Cook, 2003; Mukherji, 2006). Thus, they produce each other and have perpendicular spatial arrangement. By creating an electric field, the electrons move from the cathode to the anode producing waves. Electrical conductivity of pure epoxy system is very low which may lead to uncompleted and uneven curing as reported by Judith (1999). With addition of Carbon nanotubes (CNTs), electrical conductivity of thermosetting resins increases significantly (Gojnya et al., 2006; Allaoui et al., 2002). Furthermore, due to the very high electrical conductivity of CNTs (Bandaru, 2007), heat applies directly inside the sample which led to faster processing and obtaining absolutely cured material (Zhou and Hawley, 2003). It was found that absorption of microwave energy is associated with dipolar matrix relaxation and enhanced by the very high electrical conductivity of carbon additives (Paton and Windle, 2008; Rangari et al., 2010).

To understand the role of CNTs in microwave curing of composite, it is essential to understand microwave curing of polymers. It is possible to heat polymeric molecule because of their polar groups and segments of the dielectric material (Zong et. al, 2005). When a polymer molecule is placed in an alternating electric field, there is a change in its polarity. The energy, consumed to overcome thermal motion, is dissipated in the material and this is responsible for heating up the material. During the movement of charged electron, it displaces the charges and will polarise the material which is placed in the electromagnetic field. Displacement of charge is due to the reorientation of polar molecules (dipolar polarisation) and electronuclei dependences (Nightingale and Day, 2002). In other words, under the microwave influence, dipoles of molecules are polarised and aligned in the direction of the field (Boey and Lee, 1990; Meyer and Herbeck, 2005). Carbon based additives (including CNTs) are good absorbers of energy from microwave frequency electromagnetic fields (Zhao et al., 2006; Bao et al., 2011; Fan et al., 2006; Menendez et al.,2010). With the addition of very little amount of CNTs (0.04 wt.%), the absorbing property of composites have been increased up to 500 times as reported by Paton and Windle (2008). Zhou and Hawley (2003) reported that addition of carbon in composite materials prevents localised overheating in the microwave processing and also aids in speeding up the curing reaction. Fig. 1 describes the absorption behaviour of carbon black in epoxy resin matrix. It shows that carbon is responsible for heating up the polar groups as well as the long polymeric chain, whereas in the absence of carbon, only polar group is responsible for heating up the polymeric chain (Zhou and Hawley, 2003). Precisely, localised superheating of functional groups in polymer molecule is the main reason of curing in the absence of carbon.

Most of the research studies in this area focussed on the kinetics of microwave curing and comparisons were mainly made between the energy and time savings for conventional and microwave heating technologies. There is no research published on the effect of additions of CNT on energy and time for complete microwave curing of epoxy nancomposites. This work systematically reports the reduction in energy and time spent for microwave curing by the incorporation of various amounts of CNTs.

2. Experimental work

2.1 Material

The epoxy resin system used in this study was Araldite LY 5052/ Aradur 5052. Araldite LY 5052 is a low viscosity multifunctional epoxy system supplied by Huntsman, USA. Epoxy resin produced from bisphenol A resin and epichlorohydrin (Chernin et al., 1982). The hardener for this system was Aradur 5052 which is mixture of polyamines (Huntsman, 2010). Commercially available high purified MWNTs supplied by Electrovac, Austria (95% as per TGA, having traces of metal and metal oxide) were used as a reinforcement material. CNTs had density of 0.98 g/cm^3 (as per He pycnometery), specific surface area of 26 m^2/g, average length up to 2500 nm and average diameter of 50 nm. The synthesis method is Chemical Vapour Deposition (CVD). The mix ratio of the components was 100:33 parts by weight which correspond to 24.8 % of hardener and 75.2 % of resin, according to amine/ epoxy (A/ E) ratio due to high chemical activity of amine groups (Pinprayoon, 2007). The total mass of epoxy system was 15.9 g (3.94 g of hardener and 11.96 g of resin).

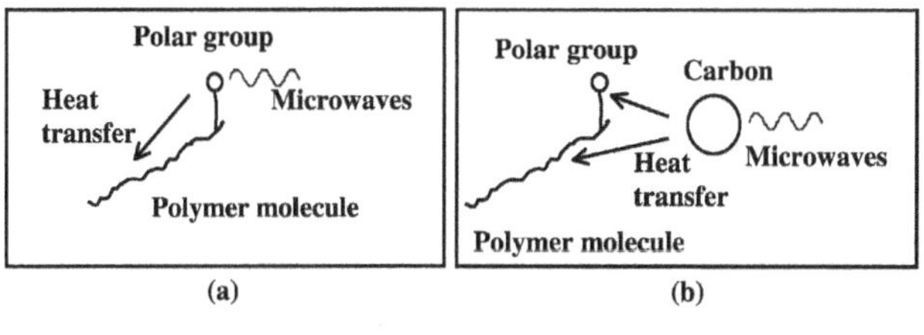

Figure 1. Microwave interactions with: a) neat epoxy; b) and epoxy doped with carbon (Zhou and Hawley, 2003).

2.2 Specimen preparation

The components of resin and hardener were weighted accurately according to the processing data and hand mixed. Pre-calculated amounts of CNTs and epoxy resin were carefully weighed and manually mixed together. MWNTs in the amount of 0.01 wt.%, 0.1 wt.% and 0.2 wt.% were infused in the matrix and dispersed via bath sonication (Ultra 7000, ultrasonic frequency: 42 kHz, power consumption: 50W) for 1 hour. Afterwards each epoxy system was divided into 6 parts and poured into the glass tubes. The tubes were put into the vacuum oven OV -11 (Medline) for 1 hour to remove the presence of air before microwave curing.

2.3 Curing procedure

The microwave setup used in this study was MARS 6 supplied by CEM Corporation, USA (magnetron frequency 2.45 GHz, power output 1800W). It was used with vessels having self-regulating control of the temperature and pressure. MARS 6 automatically recognises the type and number of vessels that have been loaded, and adjusts the output power and other parameters. The number of glass tube used in this study was 6 (maximum number of vessels is 24). Neat epoxy resin (Epoxy 1) and epoxy system infused with 0.01 wt.% CNTs (Epoxy 2), with 0.1 wt.% CNTs (Epoxy 3), and 0.2 wt.% CNTs (Epoxy 4) were cured under the same conditions. The initial parameters used are: ramp time: 10 min, hold time: 1 min, temperature: 40°C and maximum power: 500 W.

2.4 Differential scanning calorimetry

Differential Scanning Calorimetry (DSC) of various specimens was carried out with using Perkin Elmer Pyris 1 apparatus. The samples were cut into small pieces weighing from 5 mg to 10 mg using a engineering blade machine Labotom-3 supplied by Struers, Australia. Specimens were placed in aluminium plates containing a crimped lid with a small hole. The hole is necessary to maintain the constant pressure in the system and prevent deformation or rupture of aluminium pans. The DSC measurements were carried out from 30°C to 250°C at a high heating rate 10°C/ min for three cycles under nitrogen atmosphere. DSC was performed for the neat epoxy and all nanocomposite systems under the same conditions.

3. Results and discussion

Fig. 2 shows the power profile as a function of time obtained from the microwave curing process. The power profile has the same characteristics for the epoxy and nanocomposite samples during the curing process. Three stages in Fig. 2 can be visualised. Stage 1 (up to 12 sec) of the graph is a sharp decrease in the power. As shown in

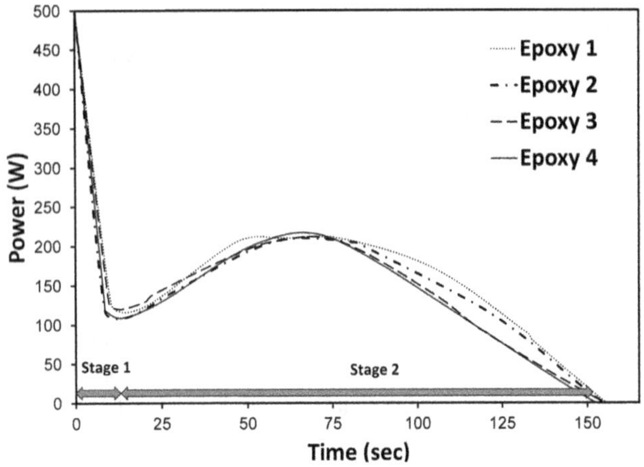

Figure 2. Power profile for the samples obtained using microwave curing.

Fig. 2 during this stage the power decreased rapidly as the default initial power (500 W) was very high for the systems. The fall of the curve extends until about a certain value of consumed power, approximately 110-120 W. Afterwards, the epoxy and nanocomposite systems start to react and absorb the heat from outside. As can be seen from Fig. 2, the chemical reaction occurred over a small period of time, from 12 sec to 154 sec, which would be referred as stage 2. The epoxy-amine reaction is the most dominated reaction during microwave curing process (Wallace et al., 2006; Mezzenga et al., 2002), whereas the epoxy-hydroxyl groups are more dominated while thermal heating. The reaction can only take place from a certain minimal energy of the incoming particle called the energy threshold of the reaction (Mezzenga et al., 2002). In this study, it can be considered that energy threshold occurs at 12 sec. Furthermore, an active chemical reaction started from 12 sec of the curing process. During stage 2, the microwave energy is absorbed by epoxy system and maximum power consumption was found to be around 220 W. At the third stage of curing (after 154 sec), in all cases, there is no energy consumption observed (Fig. 2). The epoxy resin system consumed all necessary microwave power and no external energy was required for curing. Once the required amount of heat has been absorbed, samples were cured completely (as found later by DSC analysis) and the absorption of heat was no longer detected (Fig. 2).

The area under the power-time curve (Fig. 2) was calculated by integration. Theoretically, it defines the amount of potential energy absorbed during the microwave processing. The obtained values are presented in

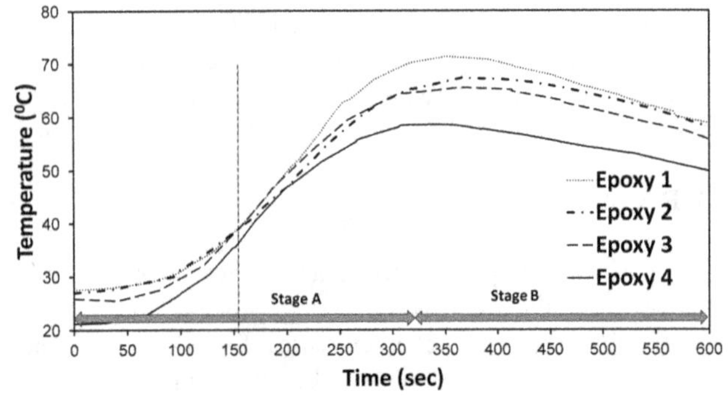

Figure 3. Temperature profile for epoxy and nanocomposite samples. The dotted line shows the point where the energy consumption was 0W.

Table 1. Summary table for microwave cured samples.

Epoxy	wt.% of CNTs	Energy, J	Maximum temperature reached during curing, °C
Epoxy 1		416	71.3
Epoxy 2	0.01	390	67.4
Epoxy 3	0.1	380	65.5
Epoxy 4	0.2	365	58.5

Tab. 1. This would indicate the amount of energy consumed for curing epoxy samples with and wihout CNTs. It was found that with the smaller addition of CNTs, more energy is required for microwave curing. The area under the curve was evaluated and compared. Conducted experiments allowed to obtain 12.5 % of energy reduction by employment 0.2 wt.% of CNTs into epoxy matrix (Tab. 1). Therefore, it can be concluded that CNTs are responsible for loweing the microwave energy consumption for curing epoxy nanocmposites.

A series of DSC test were conducted in order to observe the degree of cure for epoxy and nanocomposite samples. DSC analysis confirmed that all microwaved epoxy and nanocomposites were completely cured since there was no evidence of chemical reaction during testing. If the system is fully cured, the absorption of heat would not occur and epoxy system can be re-heated and re-cooled (PerkinElmer, 2011) reversibly below its glass transition temperature (Tg). Cyclic execution of this test is an accurate way for analysing degree of cure compared to a single heating cycle. In this work, three heating/cooling cycles were conducted prior to the reporting of results (Tab. 1). All microwaved samples were fully cured significantly before the curing time described by the supplier (Huntsman, 2010) of the epoxy system. The technical data sheet recommends curing (at room temperature) and post curing (at 100°C) time of at least 1680 min.

Fig. 4 presents the digital camera captured images of bottoms of the glass tubes with specimens after their full curing in microwave. All samples (Fig. 4) were cured using the same experimental conditions. It was observed that by the addition of CNTs, significant voids were found at the bottom of the tubes. This indicates significant heat was generated by the incorporation of CNTs during microwave curing process. Due to exothermic reaction of curing, heat was generated and as a result, gases were generated which were found trapped in the viscous epoxy melt (Fig. 4b-d). Therefore, for higher concentration of CNTs, exothermic reactions become very active which would produce larger voids in the fully cured samples as found in Fig. 4d. That also mean that lower concentra-

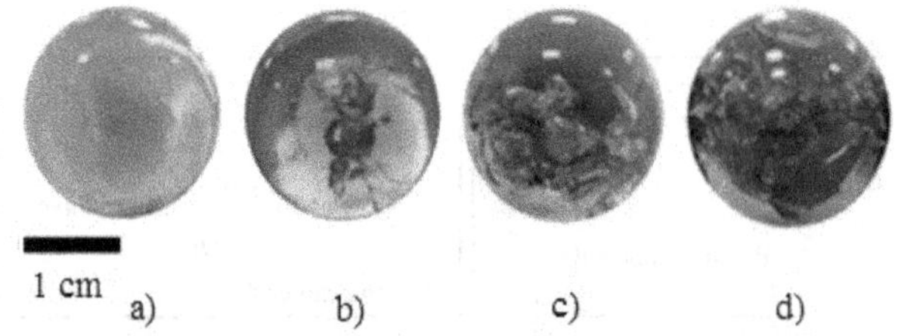

Figure 4. Digital camera captured images of bottoms of the microwaved glass tubes showing: a) neat Epoxy 1; b) Epoxy 2; c) Epoxy 3; and d) Epoxy 4.

tions of CNTs would be enough for the complete cure CNT based epoxy nanocomposites.

5. Conclusion

In this work the effect of CNT addition on the microwave curing of epoxy resin composites was investigated. The results proved that CNTs infusion can further speed up the microwave curing of CNT filled epoxy nanocomposites. CNT filled epoxy nanocomposites can be microwave cured in few minutes with significantly reduced energy consumption compared to monolithic epoxy resin without CNTs. The conducted experiments allowed obtaining 12.5 % of energy and 20.5 % temperature reductions by adding 0.2 wt. % of CNTs in epoxy matrix. DSC analysis confirmed that microwaved CNT filled epoxy nanocomposites were completely cured. This could be attributed to the good electrical and thermal conductivity and microwave absorbing properties of CNTs.

References

Allaoui, A., Baia, S., Cheng, H. and Baia, J. (2002). Mechanical and electrical properties of a MWNT/epoxy composite. Composites Science and Technology, Vol. 62, pp. 1993-1998.

Bandaru, P.R. (2007). Electrical properties and applications of carbon nanotube structures. *Journal of Nanoscience and Nanotechnology*, Vol.7, No 4-5, pp. 1-29.

Bao, T.J., Zhao, Y., Chen, L. and Duan, Y.X. (2011). Preparation and electromagnetic properties of multiwalled carbon nanotubes buckypaper/ epoxy resin nanocomposites. *Proceedings of the 18th International Conference on Composite Materials*, Jeju, South Korea.

Boey, F. and Lee, W. (1990). Microwave radiation curing of a thermosetting composite. *Journal of Materials Science Letters*, Vol. 9, No 10, pp. 1172-1173.

Chang, J., Liang, G., Gu, A., Cai, S. and Yuan, L. (2012). The production of carbon nanotube/ epoxy composites with a very high dielectric constant and low dielectric loss by microwave curing. *Carbon*, Vol. 50, No 2, pp. 689-698.

Chernin, I.Z., Smehov, F.M. and Zherdev, J.V. (1982). Epoxy resins and compositions. Moscow: Khimiya.

Cook, D. (2003). The theory of the electromagnetic field. Mineola: Dover Publications.

Das, S., Mukhopadhyay, A.K., Datta, S. and Basu, D. (2009). Prospects of microwave processing: An overview. *Bulletin of Materials Science*, Vol. 32, No 1, pp. 1-13.

Fan, Z., Luo, G., Zhang, Z., Zhou, L. and Wei, F. (2006). Electromagnetic and microwave absorbing properties of multi-walled carbon nanotubes/ polymer composites. Materials Science and Engineering B, Vol. 132, pp. 85-89.

Ghosh, N. N. and Palmese, G.R. (2005). Electron-beam curing of epoxy resins: effect of alcohols on cationic polymerization. *Bulletin of Materials Science*, Vol. 28, No 6, pp. 603-608.

Gojnya, F., Wichmanna, M., Fiedlera, B., Kinlochb, I., Bauhoferc, W., Windleb, A. and Schultea, K. (2006). Evaluation and identification of electrical and thermal conduction mechanisms in carbon nanotube/ epoxy composites. Polymer, Vol. 47, No 6, pp. 2036-2045.

Huntsman (2010). Araldite LY 5052/ Aradur 5052. Cold curing epoxy systems. [online]. Available at http://www.chemcenters.com/images/suppliers/169257/Araldite%20LY5052,%20Aradur%205052.pdf, Accessed 3 January 2013.

Judith, L. H. (1999). *Comparison of thermal and microwave processing of polyester resins.* M.Sc. Thesis, University of Manchester, Manchester.

Ku, H. S.-L. and Yusaf, T. (2008). Processing of composites using variable and fixed frequency microwave facilities. *Progress In Electromagnetics Research B*, Vol. 5, pp. 185-205.

Menéndez, J., Arenillas, A., Fidalgo, B., Fernández, Y., Zubizarreta, L., Calvo, E. and Bermúdez, J. (2010). Microwave heating processes involving carbon materials. *Fuel Processing Technology*, Vol. 91, pp. 1-8.

Meyer, M. and Herbeck, L. (2005). Microwave effects on CFRP processing. 26th *SAMPE Europe Conference & Exhibition*, JEC, Paris, France.

Mezzenga, R., Page, S. and Manson, J.A. (2002). Enthalpic, entropic, and square gradient contributions to the surface energetics of amine-cured epoxy systems. *Journal of Colloid and Interface Science*, Vol. 250, No 1, pp. 121-127.

Mukherji, U. (2006). Electromagnetic field theory and wave propagation. Oxford: Alpha Science International.

Nightingale, C. and Day, R. (2002). Flexural and interlaminar shear strength of carbon fiber/ epoxy composites cured thermally and with microwave radiation. *Composites*, Vol. 33, Part A, pp. 1021-1030.

Papargyris, D., Day, R., Nesbitt,A. and Bakavos, D. (2008). Comparison of the mechanical and physical properties of a carbon fibre epoxy composite manufactured by resin transfer moulding using conventional and microwave heating. *Composites Science and Technology*, Vol. 68, No 7-8, pp. 1854–1861.

Paton, K.R. and Windle A.H. (2008). Efficient microwave energy absorption by carbon nanotubes. *Carbon*, Vol. 46, No 14, pp. 1935-1941.

PerkinElmer. (2011). *Thermal Analysis in F1 Composites and Adhesives.* [Presentation]. Advance composite training and development centre, Hawarden.

Pinprayoon, O. (2007). *Microwave thermal analysis of epoxy resins.* M.Sc. Thesis, University of Manchester, Manchester.

Rangari, V., Bhuyan, M. and Jeelani, S. (2010). Microwave processing and characterization of EPON 862/CNT nanocomposites. *Materials Science and Engineering: B*, Vol. 168, No 1-3, pp. 117-121.

Sui, G., Zhong, W. and Zhang, Z. (2000). Electron beam curing of advanced composites. Journal of *Materials Science and Technology*, Vol. 16, No 6, pp. 627-630.

Wallace, M., Attwood, D., Day, R. J. and Heatley, F. (2006). Investigation of the microwave curing of the PR500 epoxy resin system. *Journal of Materials Science*, Vol. 41, No 18, pp. 5862-5869.

Wolff-Fabris, F. (2010). Electron beam curing of composites. Munich: Hanser Publications.

Yusoff, R., Aroua, M. K., Nesbitt, A. and Day, R. J. (2007). Curing of polymeric composites using microwave resin transfer moulding (RTM). Journal of Engineering Science and Technology, Vol. 2, No 2, pp. 151 – 163.

Zhang, J. H. and Dai, Y. C. (2006). Microwave curing and its application to aircraft structure repair. *Key Engineering Materials*, Vol.326-328, Part 2, pp. 1725-1728.

Zhao, N., Zou, T., Shi, C., Li, J. and Guo, W. (2006). Microwave absorbing properties of activated carbon-fiber felt screens (vertical-arranged carbon fibers)/ epoxy resin composites. Materials Science and Engineering B, Vol. 127, pp. 207-211.

Zong, L., Kempel, L. and Hawley, M. (2005). Dielectric studies of three epoxy resin systems during microwave cure. Polymer, Vol. 46, No 8, pp. 2638–2645.

Zhou, S. and Hawley, M. (2003). A study of microwave reaction rate enhancement effect in adhesive bonding of polymers and composites. *Composite Structures*, Vol. 61, No 4, pp. 303-309.

Zhou, J., Shi, C., Mei, B., Yuan, R. and Fu, Z. (2003). Research on the technology and mechanical properties of the microwave processing of polymer. *Journal of Materials Processing Technology*, Vol. 137, pp. 156-158.

The Analysis of Ultrajet Hydrodiagnosing Application Possibilities for Maintenance of Industrial Safety

Vladimir Tarasov, Andrey Galinovskiy, Mikhail Abashin

Space Rocket Technology Department, Faculty of Special Machinery, Bauman Moscow State Technical University,
5 2nd Baumanskaya Street, Moscow, 105005, Russia

Abstract: This paper analyses the possible applications of ultrajet technologies in mechanical engineering. Conclusions were drawn about the possible potential operational and technological parameters of liquid ultrastream usage as a control device of various objects' surface conditions. It's also concluded by the theoretical and experimental research results that show that ultrajet troubleshooting usage for industrial safety maintenance has many possibilities.

Key Words: Hydrodiagnosis, Ultrajet Hydrotechnologies, Industrial Safety, Ultrajet.

1. Introduction

It was McCartney Manufacturing's firm representatives (USA) who first took out a patent (Pat. 2985050 USA. ICP B26F 3/00. Liquid cutting of hard materials, B. C. Schwacha (USA). North American Aviation Inc., No 766952) for materials hydrojet processing and successfully applied it at Alton Box Board Co. factory for materials cutting in 1971 (Kovshov et al., 2007). Study and research works in the area of ultrajet hydrotechnologies (UJT) in the USSR began in the second half of the last century (Bilik, 1960; Vereschagin et al., 1956). The usage availability, technical and economical efficiency of given technologies is noted in the works of domestic and foreign scientists: R.A. Tikhomirov, V.M. Stepanov, V.A. Brenner, A.B. Zhabin, A.E. Pushkarev, M.M. Shchegolevsky, E.N. Petuhov, D.V. Kravchenko, G.V. Barsukov, V.A. Potapov, S.N. Poljansky, V.V. Rozanov, A.E. Provolotsky, R.A. Kuzmin, A.A. Barzov, V.S. Puzakov, V.N. Bernadsky, G. Zhou, D. Arola, D. Summers, A. Momber, M. Hashishi, M. Hessling, H. Blickwedel, J. Zeng, Y. Zhang, H. Louis, etc.

At present UJT have confidently entered into the world arsenal of the most advanced universal ways of materials physico-technical processing, which has wide range of technological possibilities.

However, as preliminary research has shown, the physico-energy basis of this technology predetermines considerably wider sphere of its practical appendices in leading industries and commercial activity.

Therefore, UJT can be defined as a complex of means and methods for the such high-energy liquid compact stream parameters creation and realization which is capable to lead to fixed purposeful changes in a processed material and/or in the liquid during its interaction with environment (for example, during the dispatch-dynamic braking against a solid-state target).

It has been shown in Bauman Moscow State Technical University (BMSTU), using functional inversion of technological concepts: the cutting tool – processed material, that liquid ultrafast stream can be considered not only as a cutting tool during material hydrocutting but also as a specific processed material – the hydrotechnological environment to ultrajet activation (Barzov, et al., 2006). Besides, it has been established that liquid ultrastream is a universal control-diagnostic tool, that allows to receive an operative information about items surface conditions parameters.

2. Ultrajet technologies characteristics and application area

UJT basis is a potential energy transformation process which is made by means of intensifying the kinetic energy of a liquid stream with a minor diameter by means of purpose-made nozzle. It can be achieved by intensifying the compression of the liquid (usually it is water compression) up to ultrahigh pressure ($p \sim 400$ MPa) and releasing it through specially profiled hydro nozzles with a diameter range of b ($d_n = 0.10...0.25$ mm). In the hydro nozzle exit section the compact water stream has supersonic speed ($V \sim 800...900$ m/s) and enormous specific kinetic energy ($E_k > 100$ MJ/kg). A supersonic liquid stream's kinetic energy, when applied to material surface area, turns into mechanical work, for example, into cutting work. Therefore the liquid stream is capable of destroying nonmetal materials as well as metal materials and alloys of almost any durability which are widely used in heat-and-power engineering (Stepanov and Barsukov, 2004; Tikhomirov et al., 1987; Tikhomirov and Guenko, 1984). An illustration of existing and predicted UJT application areas are presented in Fig. 1.

According to the scheme (Fig. 1) the experimental-theoretical attempt analysis of liquid ultrastream usage possibilities as new research means and tool control of various objects surface conditions operation-technological parameters is expedient and perspective. It is highly probable that the realization of this idea will enable the creation of a highly effective ultrajet diagnostics (UJD) system. This device will help to solve a number of important questions that are connected with the residual life estimation of potentially dangerous objects of material (fuel-energy complex, transport systems, nuclear power plants, etc). In addition to that the physical essence of the given diagnostics way is invariable, since it is inseparably related with the analysis and practical usage of liquid ultrastream and firm body energetically extreme interaction mechanisms (in particular, an interaction with its hydrodestruction mechanism).

3. Ultrajet hydrodiagnosis technology

The analysis has shown that UJD enables the definition of an objects' generic and subsurface layer condition. These parameters describe their strength, fatigue and other operational properties and when connected with a constructional material damage and diagnostics device, this complete system detects the presence of material defects and any residual stress.

The essence of UJD is that the impact of a liquid or abrasive-liquid stream onto a monitored (diagnosed) objects' explicit surface area, the parameters of this impact can be estimated: blanket plastic deformation, hydroerosion products, etc. Hydroerosion intensity (weight carry-over in unit time) can therefore be defined. Further findings are compared with source data of identical jet influence on reference material. Using the difference in impact results, (diagnosed and reference material), it is possible to conclude about the generic and subsurface physical-mechanical conditions of the controlled object, whether it is a structural element, a production sample, or an observable laboratory material sample.

The most qualitative areas (less defective, with the biggest residual resource) of diagnosed construction (in particular, least loaded zones, elements, etc) can be used like transfer standard. Transfer standard can be specially created during shaping diagnosed (in the future) construction (samples-witnesses where it is easy to define physical-mechanical properties on the direct experiments basis). Transfer standard can be also cut out, for example by hydroabrasive cut, from a run construction with the subsequent repair of formed damages.

A repaired area of a construction surface, hydrocaverns (for example ferroconcrete constructions) was subjected to jet diagnosing. This area had been repaired by filling formed area with a repair compound, for the purpose of preventing any subsequent reduction of structural capacity of the controlled object.

Thereby, it is possible to generate a UJD algorithm, which consists of following basic stages:

1) Hydrodynamic influence effects on controlled object surface, which results in surface hydroerosion on certain conditions: preset stream working pressure, its diameter, diagnosed surface rational motion kinematics, etc.
2) Hydroerosion parameters (hydrocaverns geometry characteristics, material particles weight and size, etc) are compared with reference characteristics and/or with themselves on various surface patches.
3) The controlled surface area repair quality and object conditions en masse, (current and/or predictable), are determined with received comparison results difference (absolute and/or relative).

An important aspect of applying UJD technology is final process quality control implementation on the production process. Second important aspect is possibility of positive express-evaluation of blanket and subsurface layer operational data of items and industrial details, which are

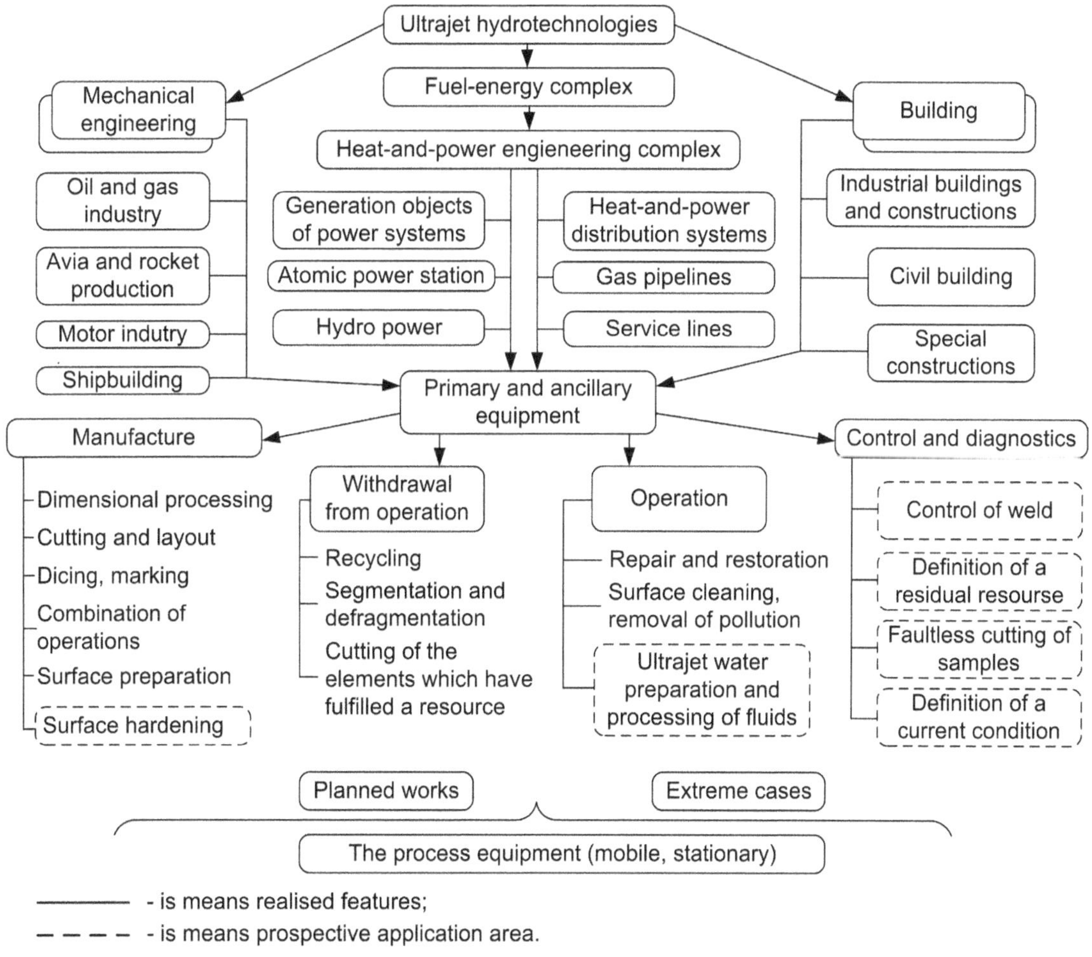

Figure 1. Ultrajet technologies application area.

the most vulnerable at producing and work. The other aspect is changing production requirements the effect of other processes and the effect of poor maintenance and external pollution on the jet systems.

Looking practically the use of UJD has the potential to increase warranty assurance on final products, for example for the possible extension of a warranties validation period, which is an actual problem of modern industry. Besides, UJD is rather useful as an operative objects diagnostics tool for materials which have been exposed to extreme conditions, for example during the estimation of structural stability after natural disaster: a fire, flooding, earthquake, etc. It is highly probable that in the near future UJD will join a number of systems that provide complicated engineering analysis of various objects projected and actual operational life.

Thereby UJD practical appendices sphere consists in information-diagnostic maintenance of all key stages of products life cycle and different function objects. UJD provides authentic diagnostic information production about state parameters, first of all about injury and its blanket and subsurface layer physical-mechanical properties, which are geometrically commensurable with hydrojet erosion geometrical parameters.

The multi-functionality of UJT (ultrajet technology), in principal, has many important practical applications. These consist of a variety of production applications, (samples reception, samples diagnostic, samples clearing, cutting and dimensional material processing, etc). Using the ultrajet equipment in a universal manner will enable the solving of a wide range of different problems, from product manufacture and repair diagnostics to assessin material and it's recycling after decommissioning from use.

The potential fields of application for UJT in fuel-energy complex should be mentioned, because ULT doesn't have alternatives in this complex but it is necessary to wait for technology development in this area. However the potential areas of research are listed below:

1. Clearing and processing of various materials without harmful and toxic gas emission: rubber, plastics, composites.
2. Polluted pipelines and communications clearing from fuel oil, petroleum and its products residues.
3. Fuel-energy complex objects cutting, segmentation and recycling with absence of processing fire-dangerous factor: tankers, petroleum pipelines, tanks-storehouses, tanks.
4. Materials cutting and dimensional processing under the water during mining operations with sea platforms usage, underwater oil pipelines laying.

4. Preliminary experimental researches results

As an example we have performed estimated results of UJD the possible of efficient express control of chemical-thermal processing, essentially raising the blanket operational characteristics at the expense of hardness increase, structure improvement, etc.

The surface, (before and after UJD case hardening), of a propulsion unit gearbox wheel of a typical gas-turbine engine, (made of a steel 16Х3МВФАБ-Ш), was studied in the experiments. Ultrajet gear wheel surface diagnostics was performed with a hydrostream with the following parameters: liquid working pressure – 400 MPa, feed – 5 mm/s, hydronozzle diameter – 0.15 mm.

The influence of Ultrastream on the gear wheel surface was monitored by means of a computer induction profilometer. This has allowed the recording of the not strengthened and cemented details surface hydrocaverns form and to receive corresponding profilogramm (Fig. 2, 3).

It follows from the received data that the strongly pronounced hydrocavern is marked on the not strengthened detail profilogramm, and strengthened detail hydrocavern parameters are within gear wheel roughness. Thereby, the findings illustrate the possibility of UJD application as means of detail hardness technological control after their chemical-thermal processing.

Another result of the experiments is in the definition of the material operational-technological properties. In the hydrocavern geometrical characteristics analysis received on samples that were azotized five, ten and fifteen hours and the sample, that was not azotized.

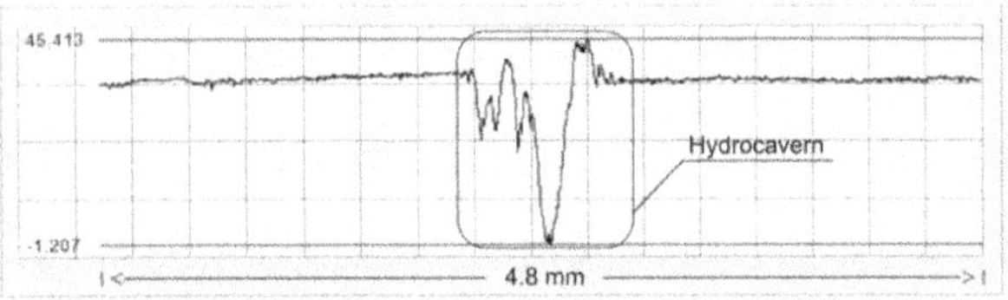

Figure 2. Hydrocavern surface profilogramm of not strengthened gear wheel (vertical axis scale is 100 micrometers/sm, horizontal axis scale is 500 micrometers/sm, gaging extent 4.8 mm)

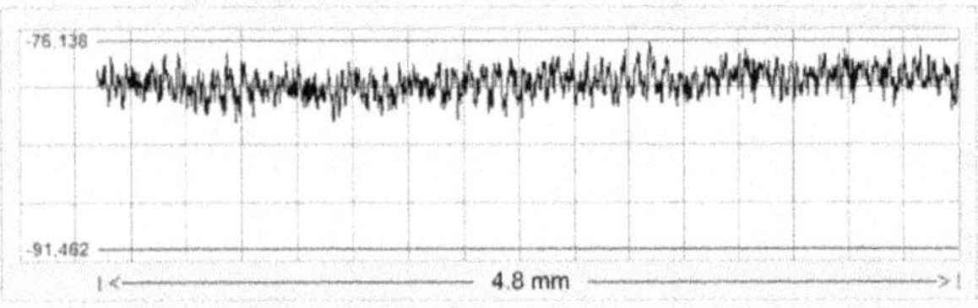

Figure 3. Strengthened gear wheel surface profilogramm (vertical axis scale is 10 micrometers/sm, horizontal axis scale is 500 micrometers/sm, gaging extent 4.8 mm).

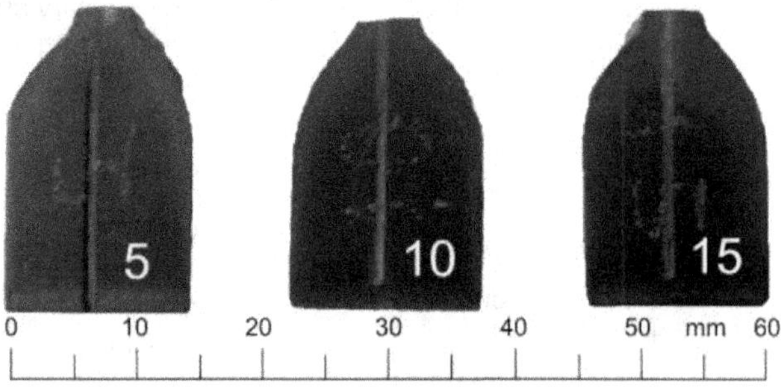

Figure 4. The nitrated samples, plunged to hydroerosive influence.

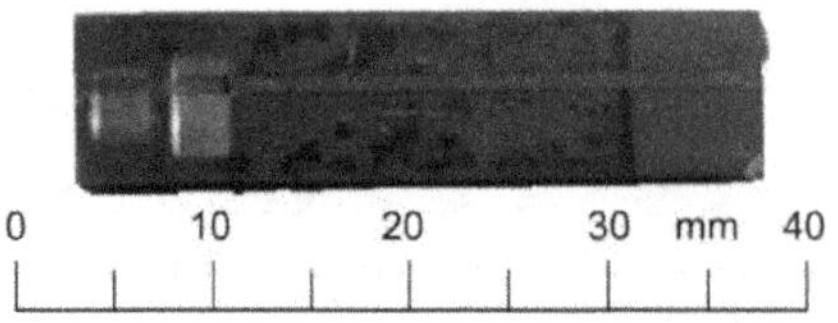

Figure 5. The sample after the wearproof tests, plunged to hydroerosive influence.

As a caverns studying result on a samples surface (Fig. 4), subjected to hydroerosive influence, some conclusions have been made – diagnosing criteria for highly rigid (in our case nitrated) samples can act on both cavity depth, and its area. Analysing the comparative study results of the samples' hydrocaverns geometrical parameters (depth) with characteristics of the nitrated layer depth have enabled the establishment of correlation values between them (0.967). This result leads to a clear conclusion about the potential application of UJD as a means of efficient operative control of various objects blanket operational-technological and quality parameters.

Similar research was conducted for the parameters of hydrocaverns using analysis received on the nitrated samples sprained wear-proof tests by means of a frictional testing machine. As a result of ultrajet hydroinfluence by an unaided sight it is possible to establish difference in hydrocavity geometrical parameters (the big width and depth) in an operative range of using up loadings and on the sample periphery (Fig. 5).

5. Conclusion

Thereby, express technique realization of details blanket quality and materials characteristics estimation provides reception of the necessary information about their condition parameters. Diagnostics is carried out by short-term (5-10 sec) influence on water ultrastream control object with the subsequent mass geometrical analysis of hydroerosive destruction products and blow zone plastic deformation character. In practice it will allow to provide necessary information about surface hardening procedure quality, to define detail residual stress level, and also to forecast an object (detail) resource as a whole. The received theoretical and experimental researches results allow asserting that UJD method has wide prospects in industrial safety maintenance decision question.

References

Kovshov, A. N., Nazarov, U. F. and Yaroslavtsev, V. M. (2007). *Netradicionnye Metody Obrabotki Materialov: The Electronic Multimedia Manual*. Moscow; MGOU.

Bilik, S. M. (1960). *Abrazivno-zhidkostnaja Obrabotka Metallov*. Moscow: GNTIML.

Vereschagin, L. F., Semerchan, A. A. and Firsov, A. I. (1956). Nekotorie issledovanija gidrodynamiki strui zhidkosti, istekauschej iz sopla pod davleniem do 1500atm. *Technical Physics Letters*, Vol. 26, No 11, pp. 2570-2577.

Barzov, A. A., Galinovsky, A. L., Puzakov, V. S. and Sydelnikov, K. E. (2006). *Ultrastrujnaja Technologia Activacii Zhidkostej*. Moscow: Mashinostroenie.

Stepanov, U. S. and Barsukov, G. V. (2004). *Sovremennie Technologicheskie Processy Mechanicheskogo i Gidrostrujnogo Raskroja Technologicheskih Tkanej*. Biblioteka technologa. Moscow: Mashinostroenie,.

Tikhomirov, R. A., Babin, V. F. and Petukhov, E. N. (1987). *Gidrorezanie Sudostroitelnyh Materialov*. Leningrad: Sudostroenie.

Tikhomirov, R. A. and Guenko, V. S. (1984). Gidrorezanie Nemetallicheskih Materialov. Leningrad: Sudostroenie.

The Preparing of the Uni-directional Glass Fiber Metal Laminates and the Fatigue Crack Propagation Characters

Zhang Jiazhen[1,2], Bai Shigang[2], Sha Yu[3]

[1] Beijing Aeronautical Science and Technology Research Institute of COMAC, Beijing 100083, China
[2] Composite Materials and Structure Research Centre, Harbin Institute of Technology, Harbin 150001, China
[3] East University of Heilongjiang, 150086, China

Abstract: The 3/2 unidirectional glass fiber metal laminates were developed using LY12-M aluminum alloy and HS2 high strength glass fiber prepreg by the hot pressing and curing progress. The fatigue crack propagation tests of LY12-M aluminum alloy plates and unidirectional glass fiber reinforced aluminum laminates were done. The result showed that the process of the preparing is feasible and the effective bridge stress of glass fiber was obtained. The fatigue crack growth rate of the laminates was 10-5~10-4mm/cycle, which was an order of magnitude lower than the LY-12 aluminum alloy. The crack opening shape was observed. The crack opening displacement of the glass fiber reinforced aluminum laminates were no significant change, however, the crack opening displacement of the single layer of LY-12 aluminum alloy increased significantly with the propagation of fatigue crack.

Key Words: Glass fiber reinforced aluminum laminate, Fatigue crack propagation, Crack opening shape.

1. Introduction

Fiber metal Laminates (FMLs) were firstly developed at Delft University of Technology as a family of hybrid materials that consist of bonded thin mental sheets and fibers embedded in epoxy (Van Lipzig, 1973). Two variants of FMLs were successively developed: Arall, containing aramid fibers, and Galare, containing glass fibers. The current investigation into fatigue crack propagation behavior focuses on Glare, which consists of aluminium 2024-T3, S2-glass fibers and the FM94 adhesive system. Glare has become known for its excellent fatigue and has been successively applied to the Airbus A380 as skin materials.

In this paper the LY-12 aluminum alloy sheet and HS2 high strength glass fiber prepreg were used for preparing of the uni-directional glass fiber metal laminates. The J272 medium temperature adhesive film was inserted between the aluminum alloy layer and glass fiber prepreg layer to enhance the interface strength.

2. Material and Methods

2.1 Material

The brand of aluminum alloy, adhesive and reinforced fiber are showed in Table 1. The Young's modulus and Yield strength are showed in Table 2.

Table 1. Materials of test

Aluminum alloy & thickness (mm)	Reinforced fiber	Adhesive
LY12-M/ 1mm, 3mm	HS2	J-272

Table 2. Material constants

Aluminum alloy		Glass fiber reinforced epoxy
Young's modulus E_{Al}/GPa	Yield strength σ_{ys}/MPa	Young's modulus E_{la}/GPa
70	120.1	45

2.2 Preparing of the uni-directional glass fiber metal laminates

The 3/2 uni-directional glass fiber metal laminates were preparing using 3 layer of LY-12 aluminium alloy sheets of 1mm thickness, 2 layer of reinforced fiber and 4 layer of adhesive film. Stacking sequence: Al/J272/Glass/J272/Al/J272/Glass/J272/Al.

The temperature of heat curing was 120~125°C. The hot pressing time was 2.0 h. The curing pressure was 0.2~0.4 MPa. The phosphoric acid anodizing was used for surface treatment of aluminium alloy sheets.

2.3. Specimens

The specimens are center cracked panels (MT). As shown in Fig. 1, the dimensions are L = 180mm, W = 40mm. The crack length is 2a. The specimens had a rectangle cross section of 10mm*1mm.

2.4 Test procedures

Fatigue test was performed on a PLG-100C high-frequency fatigue testing machine under a four baseline R ratios: R = 0.1. Test was conducted in laboratory at air

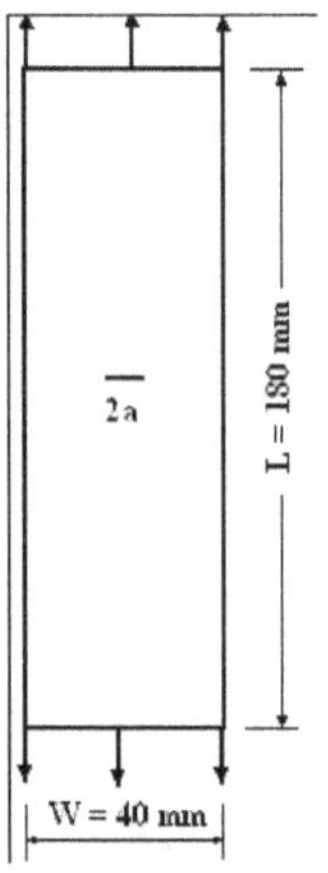

Figure 1. MT specimen geometry.

Table 3. The parameters of the experiment

Specimens	Thickness of specimens (mm)	Laminate structure	Fiber direction	Stress ratio R	Maximum tensile loading S_{max} (Mpa)	loading direction θ(°)
Laminates	3.56	3/2	0/0	0.1	57.6	0
Aluminum alloy	3.00	—	—	0.1	57.6	—

temperature of 20°C. Fig. 2 shows the high frequency fatigue testing machine and MT specimen.

The grips for the MT specimens used in the tests were specially designed such that no bending could take place and the alignment was perfect.

The maximum applied stresses of the applied stress cycle are given in Table 3. The applied maximum tensile loading is 57.6 MPa and the value remained constant. The fatigue damage in these experiments was monitored by taking photos. In this way a permanent record of the deformation and development of fatigue cracks could be recorded.

3 Test Results

3.1 The fatigue crack growth rate

The Paris law (Paris and Erdogan, 1963; Foreman et al., 1967; Walker, 1970) is

$$\frac{da}{dN} = C(\Delta K)^m \qquad (1)$$

where C and m are constants of material.

The half crack length $\{a_i\}$, the number of cycles$\{N_i\}$ data were used to calculate the fatigue crack growth rate

$$\left(\frac{da}{dN}\right)_j = \frac{a_{i+1} - a_i}{N_{i+1} - N_i}, j = i = 1, 2L \qquad (2)$$

and to calculate the stress intensity factors ΔK_j.

$$a_j = \frac{a_{i+1} + a_i}{2}, j = i = 1, 2L \qquad (3)$$

$$\Delta K_j = \frac{\Delta P}{B}\sqrt{\left(\frac{\pi\alpha_j}{2W}\right)\sec\frac{\pi\alpha_j}{2}} \qquad (4)$$

Figure 2. High frequency fatigue testing machine and MT specimen.

where, ΔP is stress amplitude $\Delta P = P_{max} - P_{mim}$, W is the width of specimens, $\alpha_j = 2a_j/W$.

The a-N curve of fatigue crack propagation of the two MT specimens was draw in the same figure to compare the performance difference of aluminum alloy and the uni -directional glass fiber aluminum alloy laminates. The Fig. 2 give the a-N curve from 2a = 6.5mm to failure. The Fig. 2 indicated that the number of cycles from 2a = 6.5mm to failure of the laminates is more than 10 times aluminum alloy.

The Fig.3 shows that the fatigue crack growth rate of the LY12-M aluminum alloy is between 10^{-4}~10^{-3}mm/cycle. Under the same stress intensity factors the fatigue crack growth rate of the uni-directional glass fiber alumi-

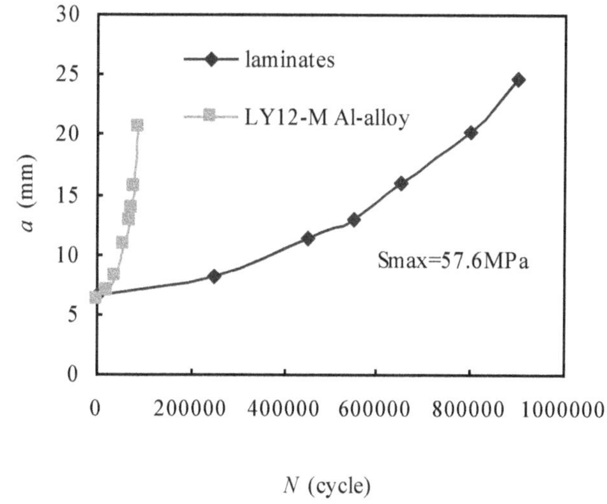

Figure 3. The a-N curve of fatigue crack propagation.

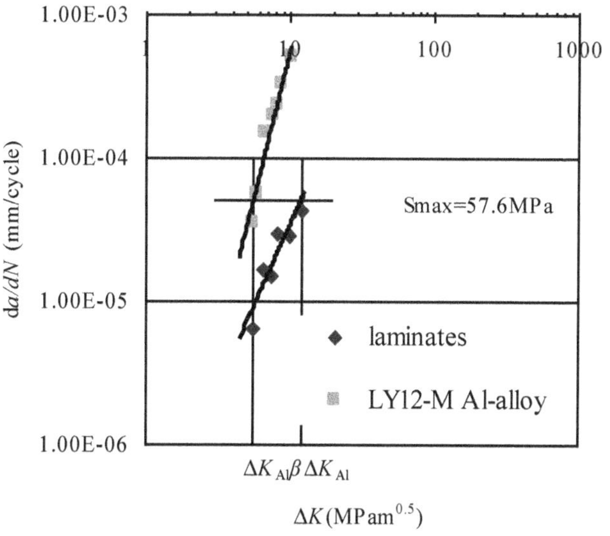

Figure 4. The da/dN-ΔK curve of fatigue crack propagation.

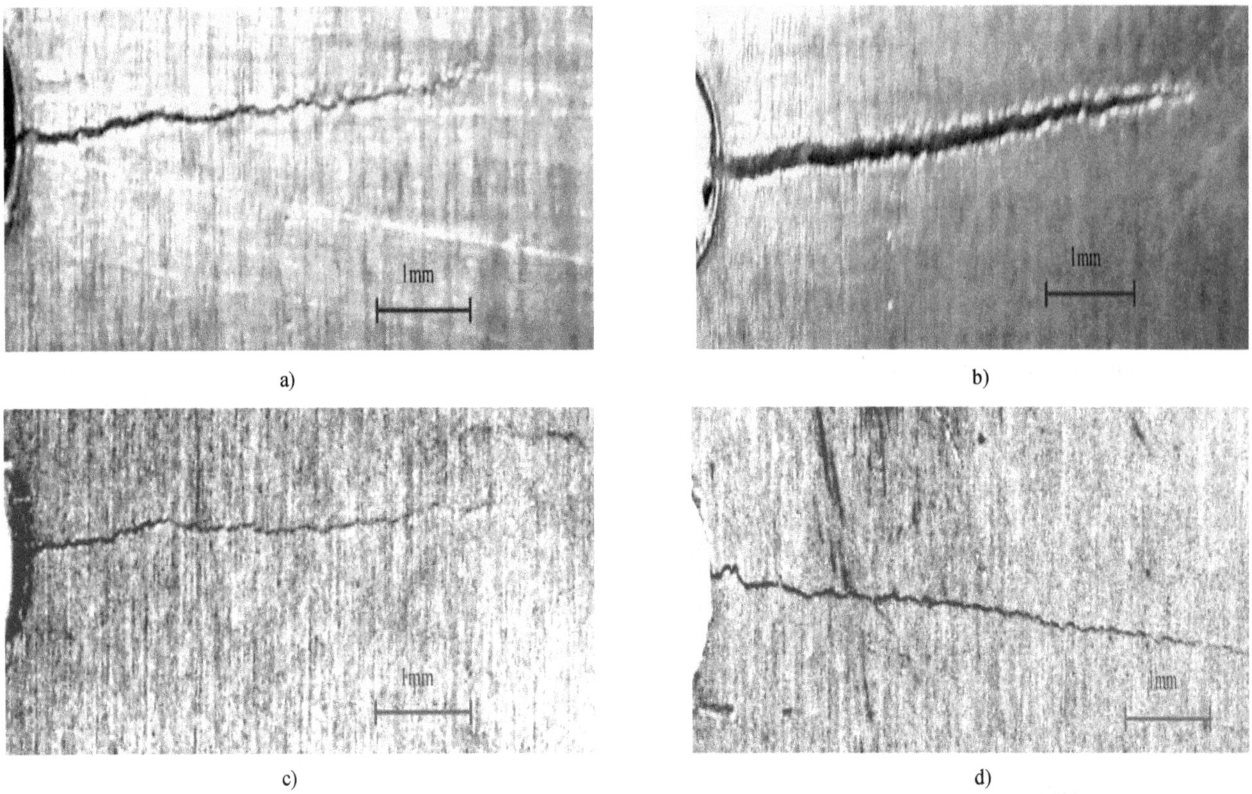

Figure 4. Crack opening shape of the single plate and laminates.

num alloy laminates is between 10^{-5}~10^{-4}mm/cycle. And the difference has increasing trend with the fatigue crack propagation.

3.2 The crack opening profile

With measuring the crack length, the crack opening profile was observed to describe the difference of fatigue property of two specimens.

Fig. 4(a) and (b) show the crack opening profiles for aluminum alloy at the maximum tension load S_{max} = 57.6 Mpa, stress ratio R = 0.1. The half crack length increased from a = 8.0mm to 11.0mm and the maximum stress intensity factor increased from K_{max} = 10.0 $Mpam^{0.5}$ to 13.0 $Mpam^{0.5}$. Fig. 4(c) and (d) show the crack opening profiles for the uni-directional glass fiber aluminum alloy laminates at the maximum tension load S_{max} = 57.6 Mpa, stress ratio R = 0.1. The half crack length increased from a = 8.0mm to 11.0mm. The maximum stress intensity factor increased from K_{max} = 10.0 $Mpam^{0.5}$ to 13.0 $Mpam^{0.5}$. It is indicated from Fig.6 that the COD (crack opening distance) of uni-directional glass fiber aluminum alloy laminates was far less than the COD of the aluminum alloy sheet. That is corresponding to the difference of the fatigue crack propagation.

3.3 Prediction of fatigue crack growth rate

Considering the bridging effect of fiber metal laminates in fiber reinforced metal laminates, the effective crack tip stress intensity factor (Guo and Wu, 1998) is the equation (5).

$$K_{FML} = K_{\max} - K_{br} \quad (5)$$

K_{br} is the bridging stress caused by stress intensity factor in equation (5). The fatigue crack growth rate of FML can be obtained by equation (1) and (5).

$$\left(\frac{da}{dN}\right)_{FML} = C\left(\Delta K_{FLM}\right)^m = C\left(\beta \Delta K\right)^m \quad (6)$$

$$\beta = F(a) \quad (7)$$

According to equation (6), equation (8)can be obtained.

$$\lg\left(\frac{da}{dN}\right)_{FML} = \lg C + m\lg\left(\Delta K_{FML}\right) \quad (8)$$

With the linear fitting data $\{\lg(da/dN)_j, \lg(\Delta K_j)\}$ of aluminum alloy LY12-M, the fatigue crack growth rate of aluminum alloy LY12-M can be concluded.

$$\lg\left(\frac{da}{dN}\right) = -\,7.4018 + 4.1366\lg\left(\Delta K\right) \quad (9)$$

$$\frac{da}{dN} = 10^{-7.4018}(\Delta K)^{4.1366} \quad (10)$$

C and m are the LY12-M aluminum alloy material constants in equation (6), C = $10^{-7.4018}$, m = 4.1366. The correlation coefficient of the fitting equation R^2, R^2 = 0.9513. It can be illustrated that the equation (9) agree well with the available test data. Fig. 2 shows β can be measured as

$$\left(\frac{da}{dN}\right)_{FML} = \left(\frac{da}{dN}\right)_{Al}$$

Fitting β and the half crack length a of equation (6), it can be got the fitting equation (11). The fitting equation

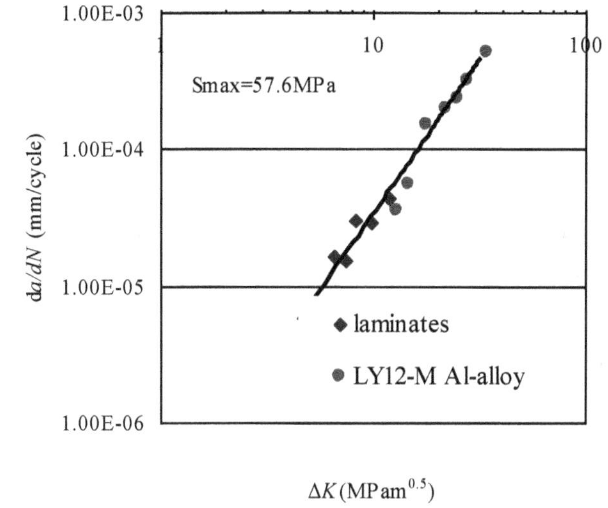

Figure 5. *da/dN* fitting line of glass fiber reinforced aluminum laminates.

R^2 = 0.9824 and the equation (11) agree well with the test data.

$$\beta = \frac{1.0908}{a^{0.3476}} \qquad (11)$$

The fatigue crack growth rate equation in the fiber direction loading of unidirectional glass fiber reinforced aluminum laminates can be concluded by equation (10) and (6). Fig. 5 shows the fitted values of equation (12) agree well with the test data.

$$\left(\frac{da}{dN}\right)_{FML} = 10^{-7.4018}\left(\frac{1.0908}{a^{0.3476}}\Delta K\right)^{4.1366} \qquad (12)$$

4. Conclusion

(1) The 3/2 unidirectional glass fiber metal laminates were developed using LY12-M aluminum alloy and *HS2* high strength glass fiber prepreg by the hot pressing and curing progress.

(2) In the range of test parameters, the fatigue crack growth rate of the LY12-M aluminum alloy is more than 10 times the one of the uni-directional glass fiber aluminum alloy laminates. The contribution factor β was imported to the Paris law to predict the fatigue crack growth rate. The β is only the function of the half crack length.

(3) With the propagation of fatigue crack, the COD of the aluminum alloy specimens increased obviously. On the contrary, the COD of the aluminum alloy laminates was very small and change slowly. The prepared aluminum alloy specimens acquired the effective bringing and enough interface strength.

Reference

Van Lipzig, H.T.M. (1973). *Retardation of Fatigue Crack Growth.* [Thesis]. Department of Aeronautical Engineering, Delft University, the Netherlands, (in Dutch)

Van Gestel, G.F.J.A. (1975). *Crack Growth in Laminated Sheet Material and in Panels with Bonded or Integral Stiffeners.* [Thesis]. Department of Aeronautical Engineering, Delft University, the Netherlands, (in Dutch)

Hoeymarkers, A.H.W. (1977). *Fatigue of Lugs.* [Thesis]. Department of Aeronautical Engineering, Delft University, the Netherlands, (in Dutch)

Roebroeks, G.H.J.J. (1991). *Towards - The Development of a Ffatigue Insensitive and Damage Tolerant Aircraft Material.* [Thesis]. Delft University of Technology.

Human, J.J. (2006). Fatigue initiation in fibre metal laminates, *International Journal Fatigue*, Vol. 28, pp. 366-374

Paris, P.C. and Erdogan, F. (1963). A critical analysis of crack propagation laws. *Trans. ASME, J. Basic. Eng.*, Vol. 85, pp. 528-534.

Foreman, R. G. et al. (1967) Numerical analysis of crack propagation in cyclic load structure. J. Basic Engineering, 89: 454-464.

Walker, K. (1970). The effect of stress ratio during crack propagaiton and fatigue for 2024-T3 and 7075-Tt aluminum. *ASTM STP 462, Effect of Environment and complex load History on Fatigue Life*, pp. 1-14.

Guo, Y.J. and Wu, X.R. (1998). A theoretical model for predicting fatigue crack growth rates in fibre-reinforced metal laminates, *Fatigue & Fracture of Engineering Materials & Structures*, Vol. 21, pp. 1133 -114.

Takamatsu, T. et al. (2003). Evaluation of fatigue crack growth behaviour of 3 fiber/metal laminates using compliance method, *Engineering Fracture Mechanic*, Vol. 70, pp. 2603-2616.

Tada, H. et al. (2000). *The Stress Analysis of Crack Handbook*, 3rd edn., The American Society of Mechanical Engineers.

Design of Temperature Measurements in Carbon Epoxy Polymer Structures under Microwave Treatment

Richard Day[1], Sergey Reznik[2], Sergey Rumyantsev[1,2]

[1] Institute for Art, Science and Technology, Glyndŵr University, Plas Coch, Mold Road, Wrexham, LL11 2AW, UK
[2] Rocket and Spacecraft Composite Structures Department, Faculty of Special Machinery, Bauman Moscow State Technical University, 5 2nd Baumanskaya Street, Moscow, 105005, Russia

Abstract: Polymer composite materials (PCM) are coming into wider use in engineering, construction, sports and consumer goods manufacturing. However, conventional PCM curing technologies possess a number of drawbacks with regard to energy efficiency and process duration. Microwave energy can be used to rapidly and efficiently heat the PCM parts. Characteristic features of microwave heating call for specialized devices of temperature measurement. A heat mathematical model is presented for a PCM workpiece fixed to a cylindrical mandrel with variously positioned temperature sensors under microwave treatment. Data on rational positioning of the sensors were obtained and interpreted; temperature measurement errors were estimated.

Key Words: Composite, Temperature, Measuring, Microwaves.

1. Introduction

Unique physical and mechanical properties of polymer composite materials account for their increasingly wider utilization in aerospace applications, as well as in other fields of engineering, construction industry, sports equipment and consumer goods manufacturing. The growing interest in PCM makes the manufacturers acutely aware of the immediate need for technological processes improvements. The primary concern is to achieve PCM curing in less time without compromising the quality. Conventional curing methods are based on convective, radiant or contact heat exchange between the surface and the heat source. However, low thermal conductivity of polymers and resulting thermal gradients in the material increase the process duration. There is a perspective technology that is getting particular attention of late: microwave treatment (Abdulsalamov et al., 1992; Morozov et al., 2011; Papargyris et al., 2008; Nightingale and Day, 2002), which intensifies curing process in PCM and improves the quality of the product.

2. Specifics of temperature measurement under microwave treatment

Microwave heating and curing of PCM offer a number of advantages over traditional heating methods. Firstly, thermal gradients in the bulk of the material can be reduced to the minimum owing to the volumetric nature of heating. This feature enables significant reduction (2 and more times) in the processing time by increasing the heating rate (Papargyris et al., 2008; Nightingale and Day, 2002). Secondly, microwave energy is mainly directed at heating of the dielectric part, not the oven itself, which results in lower energy consumption and higher level of process control (Nightingale and Day, 2002; Rumyantsev and Minakov, 2011; Reznik et al., 2012).

Technological processes for thermosetting resins require active monitoring of the part's temperature over the whole heating cycle. The majority of the contact temperature measurement devices used in conventional ovens (such as thermocouples and thermistors) to be utilized in microwave ovens need additional devices to prevent the electromagnetic field interfering with the incoming signal and heating the conductor. The solution is to use fiber optical sensors immune to microwave radiation. At the same time, sensors positioned on the surface or in the bulk of the processed part would interfere with its temperature field due to different thermal physical characteristics of their materials, while a number of closely positioned sensors will produce mutual temperature interference. It is of significant practical interest to determine the value of temperature field interference, i.e. the temperature measurement methodical error (TMME). There exist

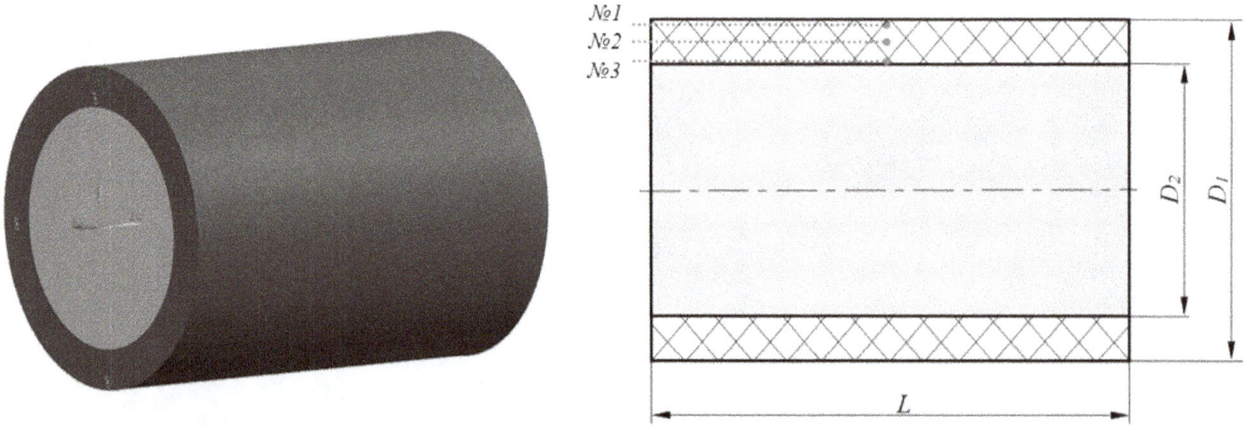

Figure 1. Geometric model of "part – mandrel – temperature sensors" system (1-3 – position of sensors).

Table 1. Properties of Hexel M21 Prepreg

T, °C	30	40	50	60	80	100	120	140	160	180
c, J/(kg·K) (across fibers)	840	820	830	850	1000	1190	1430	1550	1700	1990
λ, W/(m·K) (across fibers)	0.35	0.35	0.32	0.33	0.34	0.4	0.47	0.52	0.52	-
Emissivity	0.89									
Density, kg/m^3	1530									

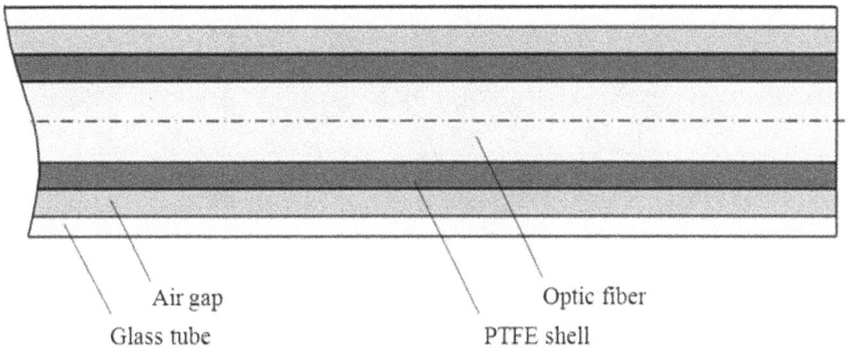

Figure 2. Fiber optic sensor model.

numerous papers on TMME analysis for various sensors layouts. J.V. Beck, V.N. Eliseev, N.A. Yaryshev and others made a great contribution to the sphere of contact temperature measurement (Beck and Hurwicz, 1960; Gerashchenko et al., 1989; Yaryshev, 1990). Early papers were mainly concerned with analytical solutions for the heat exchange in a "sensor – sample" system, with finite difference method and finite elements method coming into use at a later stage of research (Mikhalev and Reznik, 1988; Reznik et al, 2009; Reznik et al., 2012).

The object of this research is to estimate TMME for variously positioned fiber optic sensors in a hollow cylindrical part fixed to a cylindrical mandrel from an inert material under microwave treatment.

3. Heat model and conditions of estimating the temperature measurement methodical error for a microwave field processed part

The geometric model is a cylindrical shell of a workpiece tightly fixed to a solid cylindrical mandrel, so that the thermal contact between the part and the mandrel is assumed to be ideal. The part and the mandrel are of the same length. Three sensors are circumferentially at 90° positioned in the bulk of the PCM work piece at various distances from the internal surface. The model also provides for sensors positioned close to the external and internal surfaces, as well as in the bulk. The model of the fiber optic temperature sensor (Fig. 2) is a multilayered structure: the external protective cover (0.2 mm thick glass), air gap (0.2 mm thick), PTFE shell (0.2 mm thick), and a thin optic fiber (0.6 mm diameter). The sensitive element – a layer of luminescent substance over the end of the fiber – was not included into the model because of its small thickness.

The depth of penetration into the bulk of the part constituted half of the cylinder length, with a 1 mm air gap between the end of the sensor and the wall of the canal. The thermal contact between all layers of sensor and workpiece materials was assumed ideal.

The solving algorithm was based on the finite elements method implemented through Siemens NX8 and MSC Nastran software. The thermal physical properties of the modeled objects' materials are presented in Tables 1-2

Table 2. Properties of the Materials in the Model

	T, °C	30	50	100	150	180
Quartz glass	c_p, J/(kg·K)	735	750	840	890	900
	λ, W/(m·K)	1.35	1.4	1.46	1.55	1.6
	Density, kg/m^3	2200				
PTFE	c_p, J/(kg·K)	1040				
	λ, W/(m·K)	0.23	0.23	0.24	0.25	0.25
	Density, kg/m^3	2200				
Mandrel material	c_p, J/(kg·K)	1000	1010	1020	1035	1045
	λ, W/(m·K)	0.13	0.14	0.16	0.18	0.19
	Emissivity	0.89				
	Density, kg/m^3	650				

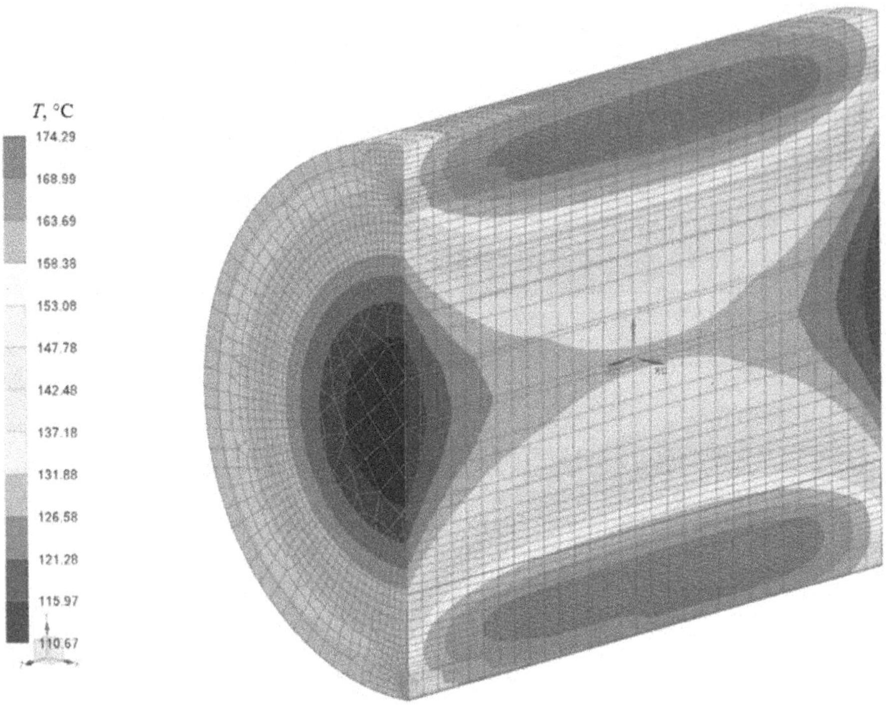

Figure 3. Temperature distribution in the heated bodies (cross-section at the 700th sec of the microwave heating).

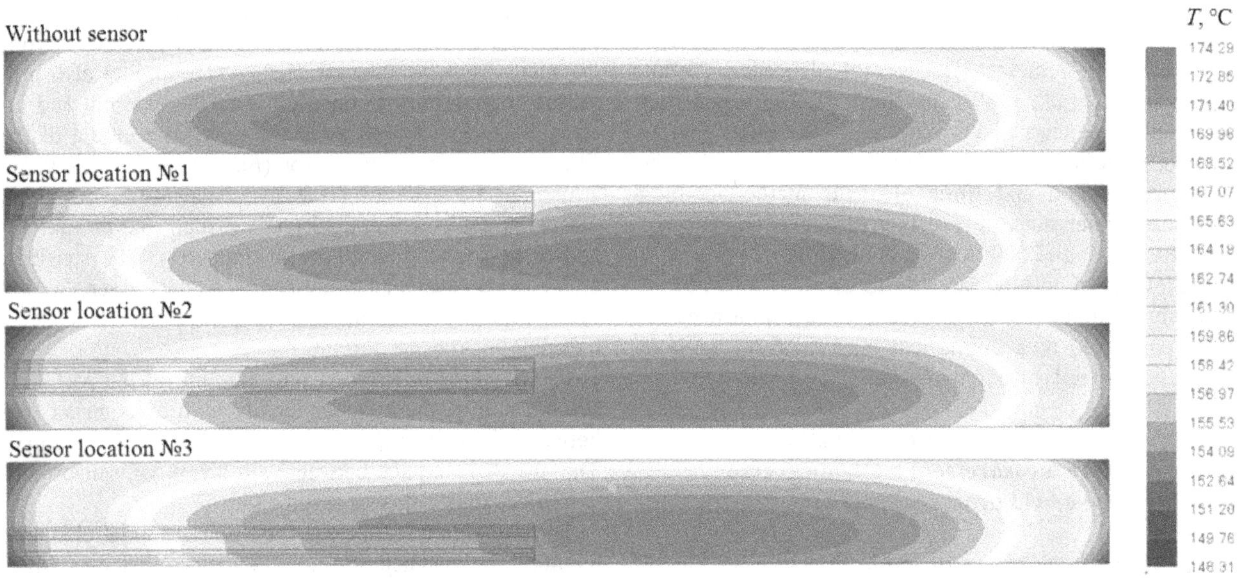

Figure 4. Comparative heat distribution across the composite workpiece section: no sensor, sensor at the outer surface (position №1), sensor on the middle surface (position №2), sensor at the inner surface (position №3).

(Kabanov, 1977; Trigorlyi, 2000; Reznik, 1982). Adaptive spatial mesh refinement was used at the sensors locations. The model utilized the finite elements of Hex8 and Wedge6 type, with the total amount of 84,510 and the total amount of nodes of 97,680.

The electromagnetic field energy absorbed by the workpiece (mean value of 85 W) was modeled as a combination of interior heat sources uniformly distributed in the bulk of the sample. The heat radiated as a result of PCM curing reaction was added to the microwave field energy. The heat from the outer surface and the ends of the workpiece and the mandrel was emitted into the environment through convection and heat radiation. The duration of the modeled treatment process constituted 700 sec, with a 2 sec interval. The heat transfer coefficient was assumed to be 3 W/(m^2·K).

4. The results of mathematical modeling and their interpretation

Mathematical modeling of the microwave treatment process indicates transient temperature field in the composite work piece. Fig. 3 shows that the temperature field is fairly uniform, with maximum gradients at the exterior surface not exceeding 5 K/mm.

In order to compare temperature field disturbances resulting from the sensor material in different positions analysis was conducted and diagrams of heat distribution in sensors locations were obtained. Fig. 4 shows the comparative effect of the variously positioned sensor on the temperature distribution in the bulk of the part.

In addition to the heat distribution diagrams, graphs were created to present the temperature changes in the

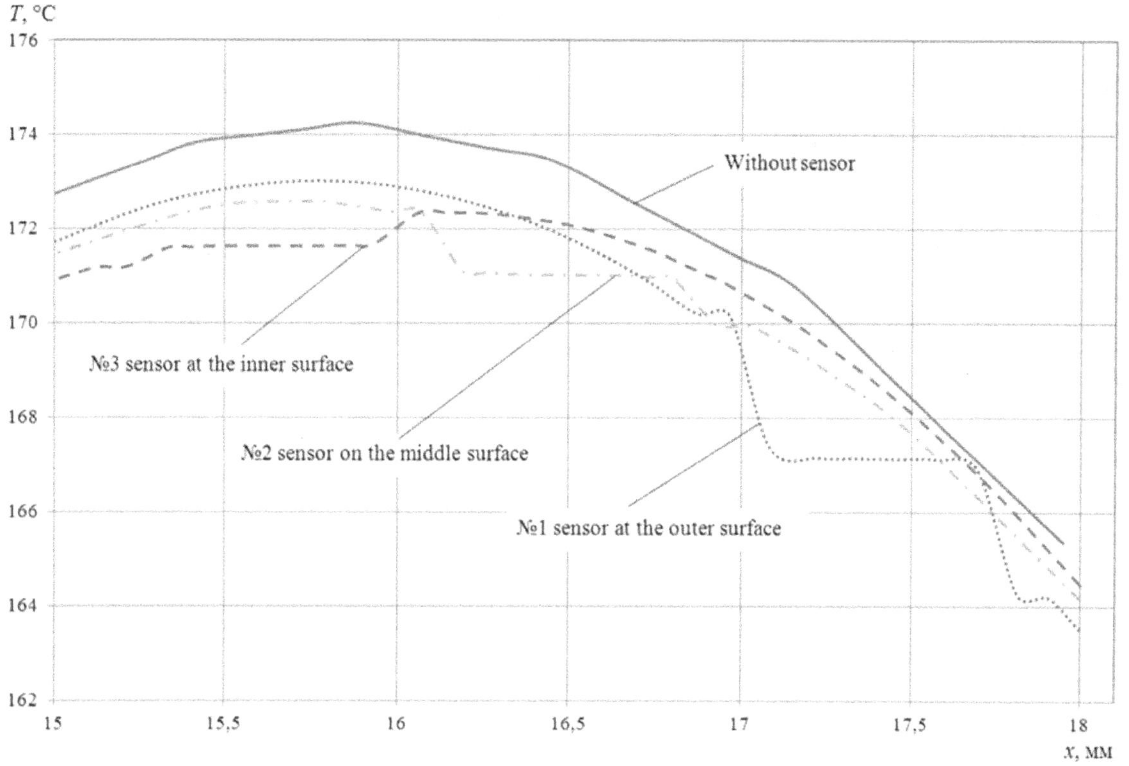

Figure 5. Temperature distribution in the L/2 work piece cross section.

bulk of the PCM work piece at the location of the sensor compared to the area without a sensor (Fig. 5). Estimates indicate that the calculated TMME value in positions 1, 2 and 3 constitutes 1.89 K, 2.41 K, 2.44 K correspondingly.

In addition to the heat distribution diagrams, graphs were created to present the temperature changes in the bulk of the PCM work piece at the location of the sensor compared to the area without a sensor (Fig. 5). Estimates indicate that the calculated TMME value in positions 1, 2 and 3 constitutes 1.89 K, 2.41 K, 2.44 K correspondingly.

5. Conclusion

Microwave heating of a cylindrical PCM with variously positioned temperature sensors was performed and interpreted. Calculation data on the temperature at various points in time were obtained, with qualitative data of the sensor material effect on the temperature field of the heated work piece. The results of the modeling having been compared, it was established that the rational location of the sensor with the minimum TMME corresponded to the position at the outer surface of the work piece (position 1). TMME for this position does not exceed 2 K, with 2.5 K for the other positions.

References

Abdulsalamov, V.M., Bezlyudova, M.M., Bulanov, I.M., Shvorobey, Yu.L. and Shvorobey, V.Yu. (1992). *Basics of the Microwave Heating of Polymer and Composite Materials*. Moscow. (in Russian)

Beck, J.V. and Hurwicz, H. (1960). Effect of thermocouple cavity on heat sink temperature, *Journal of Heat Transfer*, Vol. 82, No.1, pp. 27-36.

Geraschenko, O.A. (ed.), Gordov, A.N., Eremina, A.K. et al (1989). *Temperature Measurements*. Reference book, Kiev. (in Russian)

Kabanov, V.A. (1977). *Encyclopedia of Polymers*, Moscow: Soviet Encyclopedia. (in Russian)

Mikhalev, A.M. and Reznik, S.V. (1988). Determination of methodological error for thermocouples measurement in semitransparent scattering materials under transient heating. 1. Mathematical Model, *Izvestija VUZov: Mashinostroenije*, No 2. (in Russian)

Morozov, G.A., Morozov, O.G., Nasybullin, A.R. and Samigullin, R.R. (2011). Microwave treatment of thermosetting and thermoplastic polymers, *Physics of Wave Processes and Radio Technical Systems*, Vol. 14, No 3, pp. 114-121. (in Russian)

Nightingale, C. and Day, R.J. (2002). Flexural and interlaminar shear strength properties of carbon fibre/epoxy composites cured thermally and with microwave radiation, *Composites*, Part A: Applied science and manufacturing, No 33, pp. 1021-1030.

Papargyris, D.A., Day, R.J., Nesbitt, A. and Bakavos, D. (2008). Comparison of the mechanical and physical properties of a carbon fibre epoxy composite manufactured by resin transfer moulding using conventional and microwave heating, *Composites Science and Technology*, No 68, pp. 1854-1861.

Reznik, S.V. (1982). Experimental and theoretical determination of thermal physical properties of heat insulating materials, *Collected Papers of Bauman MVTU*, No. 392, Issues of heat transfer and thermal testing of structures, pp. 55-62. (in Russian).

Reznik, S.V., Anuchin, S.A., Prosuntsov, P.V. and Shulyakovskiy, A.V. (2009). Consideration of the procedural error for measuring contact sensor temperature during thermophysical studies, *Refractories and Industrial Ceramics*, No 3, pp. 29-33. (in Russian)

Reznik, S., Rumyantsev, S. and Guzeva, T. (2012). Heat mathematical model of dielectric composite cylinder during microwave treatment, *Proceedings of the Second International Conference on Advanced Composite Materials and Technologies for Aerospace Applications*, ACMTAA-2012, June 11-13, 2012, Wrexham, North Wales, UK, pp. 87-91.

Reznik, S.V., Sereda, G.N. and Shulakovskiy, A.V. (2011). The method of measurement of temperature of missile fairings by thermocouples in bench thermal tests, *Thermal Processes in Engineering*, Vol. 3, No 6, pp. 278-288. (in Russian)

Rumyantsev, S. and Minakov, D. (2011). Estimation analysis of temperature distribution in composite material samples under SHF-waves treatment, *Proceedings of the First International Workshop on Advanced Composite Materials and Technologies for Aerospace Applications*, ACMTAA-2011, May 9-11, 2011, Wrexham, UK, pp. 149-157.

Trigorlyi, S.V. (2000). Numerical modeling and process optimization of ultrafrequency heat treatment of dielectrics, *Applied Mechanics and Technical Physics*, Vol.41, No 1, pp. 112-119. (in Russian)

Yaryshev, N.A. (1990). *Theoretical Basics of Transient Temperatures Measurements*, Leningrad. (in Russian)

A Review of Finite Element Modelling of Composite Structures Subjected to Blast Loads

Zhongwei Guan

School of Engineering, University of Liverpool, Brownlow Street, Liverpool, L69 3GQ, United Kingdom

Abstract: This paper presents a brief review of numerical modelling of composite laminates and sandwich structures subjected to blast loading. In blast, structures undergo large plastic deformations or failure within a very short time duration. The review covers different approaches to simplify impulsive loading, available constitutive relationships, failure criteria and damage evolution laws in modelling of various constituent materials in composites. Although extensive work has been undertaken on modelling blast responses of various composite structures, capture of the perforation failure characteristics still remains as a challenging task. More advanced failure criteria and damage evolution mechanisms are required to tackle this problem.

Key Words: Blast, Finite element, Composite laminates, Sandwich structures.

1. Introduction

With more uncertainties in the modern world due to possible terrorist threats or natural disasters, the blast response and resistance of composite structures become increasingly important issues to engineering communities and government bodies. In the initial stage of the explosion, an impulsive pressure, much greater than the static collapse pressure of the structures, is produced by shock wave in a very short time period. This extremely high pressure forces the structures undergo large plastic deformations and collapse or perforation failure, associated with energy absorption. In order to enhance energy absorption and blast resistance performance of composite structures, a deep insight into the explosion loading characteristics and the relationship between such the load and the structural deformation/failure behaviour is required.

Testing the blast response of the composite structures is very time consuming, in addition to material and manpower costs. When conducting experimentally-based research, tests should cover as many scenarios as possible, such as stacking configurations, materials, geometries, loading and boundary conditions, etc. Therefore, the optimisation process is likely to be very expensive and hugely time consuming. In contrast, the development of computer models using finite element analyse is a relatively quick and inexpensive process. In such circumstances, only a limited number of material tests and structural blast tests are required for validation purposes. Once computer models are verified against typical blast tests, covering both the upper and lower bound cases, systematically-designed parametric studies can be undertaken using validated numerical models.

Although reasonably extensive work has been undertaken on modelling blast responses of various composite structures, capture of the perforation failure characteristics still remains as a challenging task. More advanced failure criteria and damage evolution mechanisms are required to tackle this problem. In this paper, review will be focused on numerically describing blast loading with acceptable simplifications, constitutive models, failure criteria and damage evolution laws used to simulate the blast response of the following composite structures. These cover fibre metal laminates, carbon and glass fibre reinforced laminates. The sandwich structures reviewed include PVC foam core based with composite skins or metal skins, metallic foam core with metallic skins and metallic cellular core with metallic skins.

2. Characterising and modelling blast loads

To model the blast behaviour of composite laminates and sandwich structures, the mechanisms of blast pressure building-up after detonation need to be modelled in the first instance. Blast is an extreme dynamic loading process. In the process of an explosion, the blast travels as an incident wave until it strikes an object. Upon striking the object, a reflected wave is generated which travels back towards the point of explosion. At a point, some distance from the centre of the explosion, the reflected wave meets the incident wave, producing a single vertical wave front. The structure below the point of intersection of the reflected wave and the incident wave experiences a single shock, whereas the surface above this point experiences a shock history, which is a resultant of the incident and reflected wave. During the blast processes, the pressure builds up to a peak value of an excessive pressure. The pressure then decays to a local ambient pressure in a time point to a partial vacuum of very small amplitude and eventually returns to the ambient pressure. The portion of the pressure-time history below zero is referred to as the "negative or suction phase" and the portion above zero is called the "positive phase", which has a little effect on the response of structures (Brode, 1955). Fig. 1 shows a typical pressure-time history for a blast wave (Kinney and Graham, 1985).

In most blast studies, the negative phase of the blast wave is ignored, and only the parameters associated with the positive phase are considered since it is generally accepted that the damage to the structure is caused by the positive phase (Xue and Hutchinson, 2003; Qiu et al., 2003). The positive impulse of the wave during the positive phase (Fig. 1) can be give as

$$I = \int_{t_a}^{t_a+t_d} P(t)dt \qquad (1)$$

where $P(t)$ is the positive pressure as a function of time.

There are different forms of $P(t)$ proposed by research-

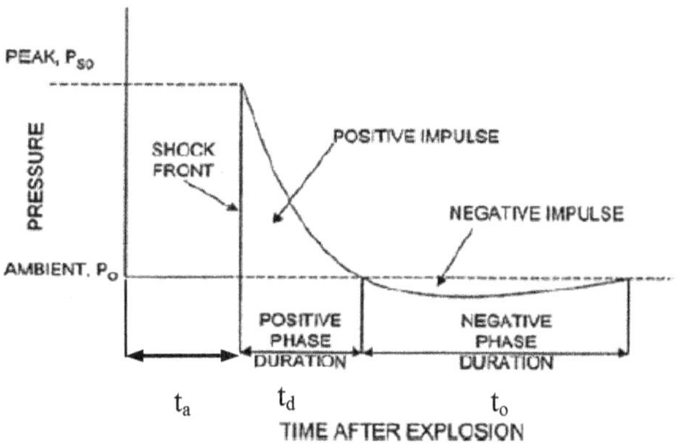

Figure 1. Blast wave pressure – time history showing positive and negative phase durations.

ers, which is often described by exponential functions such as the Friedlander equation (Kinney, 1962; Smith and Hetherington, 1994)

$$P(t) = P_{SO}\left[1-\frac{t}{t_d}\right]\exp\left\{-\frac{bt}{t_d}\right\} \tag{2}$$

In the equation b is the decay factor which is a function of P_{so} (Fig. 1). The pressure-time profile is shown in Fig. 2.

The simplified profile was also used by researchers (Qiu et al., 2004; Mori et al., 2007; Batra and Hassan, 2007; Batra and Hassan, 2008), which is expressed as

$$P(t) = P_{SO}e^{-t/\lambda} \tag{3}$$

where λ is the decay constant. The pressure-time profile in Eq. (3) can be shown in Fig. 3. A triangular blast and Heaviside pulse was also proposed (Marzocca et al., 2001) and used (Librescu et al., 2004), i.e.

$$P(t) = P_{SO}\left[1-\frac{t}{t_d}\right]\left[H(t)-\delta_b H(t-t^* t_d)\right] \tag{4}$$

In the above equation, $H(t)$ denotes the Heaviside step function, δ_b is a tracer that can be taken as 1 or 0 depending on whether the sonic-boom or triangular blast load is considered, t^* denotes the shock pulse length factor. Based on Friedlander equation in Eq. (2), a modified Friedlander equation is proposed in Eq. (5) below.

$$P(t) = P_{SO}\left[1-\frac{t-t_a}{t_d}\right]\exp\left[-A\left(\frac{t-t_d}{t_d}\right)\right] \tag{5}$$

where A is a decay coefficient.

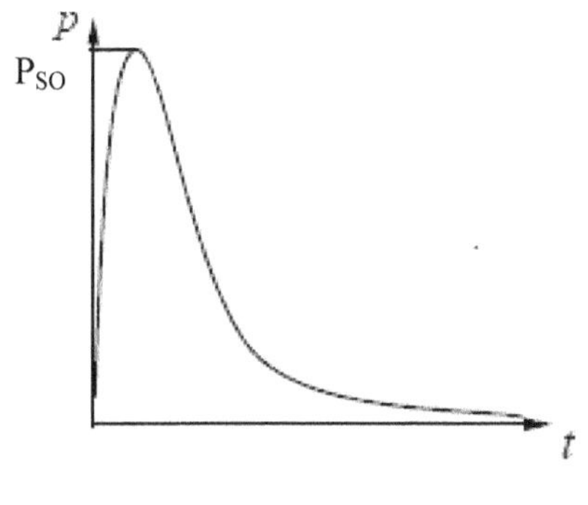

Figure 2. A schematic diagram of a pressure-time pulse.

In numerical modelling of blast load, apart from the pressure-time profile, pressure-location relation also needs to be considered. For a square laminated plate subjected to surface explosion, the pressure function $P(r,t)$ was proposed (Bonorchis and Nurick, 2009) and the pressure pulse shape was implemented with AUTODYN (Karagiozova et al., 2010).

$$P(r,t) = p_1(r)p_2(t) \tag{6}$$

where

$$p_1(r) = \begin{cases} P_{SO} & r \le r_0 \\ P_{SO}e^{-k(r-r_0)} & r_0 < r < r_b \\ 0 & r > r_b \end{cases} \tag{7}$$

$$p_2(t) = e^{-2t/t_d}$$

In Eq.(7), r_0 is the radius of the explosive disc used in the experiments (Landon et al., 2007), k is an exponential decay constant, which models the pressure distribution over the exposed area of the plate, $r_b < L/2$, L being the length of the panel. The decay constant $k(I)$ with the variation of the total impulse, I, used in the numerical simulations can be interpolated. The total impulse is defined as

$$I = 2\pi\int_0^{\infty}\int_0^{r_b} P(r,t)\,r\,dr\,dt \tag{8}$$

Vo et al. (2012) implemented both Eqs. (6) and (7) into ABAQUS/Explicit through a user-defined subroutine. The less computational expensive ConWep algorithm (Hyde, 1992) used to describe air blast loading is also available in LS-DYNA and ABAQUS/Explicit. Other simplified or idealised pressure-time profiles were proposed (Rajamani and Prabhakaran, 1980; Thiagarajan et al., 2003).

3. Modelling of composite laminates and sandwich structures

Composite structures subjected to blast loading can be modelled based on modelling blast response of their constituent materials and bonding between them. Such structures are made of monolithic materials or cellular solids. There are a number of constitutive relationships, failure criteria and damage evolution mechanisms available to determine their corresponding dynamic behaviour under blast loading conditions. In addition, several contact algorithms are also available to model interaction between constituent materials.

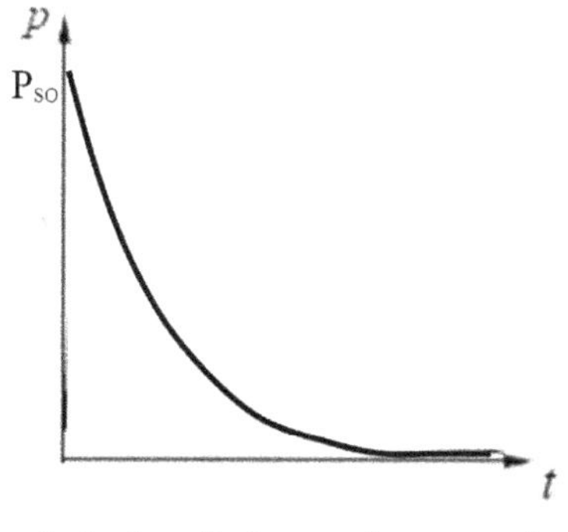

Figure 3. A schematic diagram of a pressure-time pulse expressed by Eq. (3).

3.1 Modelling metallic materials

There are three commonly used material models for metals subjected to high strain rate loading. The Johnson-Cook constitutive model is perhaps the most widely used one at the present, which can be used to describe metal plasticity under large strain, high strain rate and temperature conditions (Johnson and Cook, 1983; Johnson and Cook, 1985).

$$\bar{\sigma} = \left[A + B(\bar{\varepsilon}_{pl})^n\right]\left[1 + C\ln\left(\frac{\dot{\bar{\varepsilon}}_{pl}}{\dot{\varepsilon}_o}\right)\right]\left[1 - \left(\frac{T - T_{room}}{T_{melt} - T_{room}}\right)^m\right] \quad (9)$$

where $\bar{\sigma}$ is the equivalent stress, $\bar{\varepsilon}_{pl}$ is the equivalent plastic strain, n is a strain hardening index, $\dot{\bar{\varepsilon}}_{pl}$ is the equivalent plastic strain rate, $\dot{\varepsilon}_o$ is the reference strain rate, T_{melt} and T_{room} are the melting and transitive temperatures respectively and A, B, C and m are material constants.

The Johnson-Cook damage law (Johnson and Cook, 1985) can be used to simulate failure of the metals. This model is given as:

$$\varepsilon_f = \left[D_1 + D_2\exp\left(D_3\frac{p}{\bar{\sigma}}\right)\right]\left[1 + D_4\ln\left(\frac{\dot{\bar{\varepsilon}}_{pl}}{\dot{\varepsilon}_o}\right)\right] \times \times\left[1 + D_5\left(\frac{T - T_{room}}{T_{melt} - T_{room}}\right)\right] \quad (10)$$

where ε_f is the equivalent strain to fracture at the current conditions of strain rate, temperature, pressure and equivalent stress, p is a pressure stress, $D_1 - D_5$ are damage parameters.

For rate dependent metallic materials, the uniaxial flow rate is expressed as the following relationship.

$$\dot{\varepsilon}_{pl} = h(\bar{\sigma}, \bar{\varepsilon}_{pl}, T) \quad (11)$$

where h is a known function. The rate-dependent hardening curves in terms of the static relation can be expressed as (Stout and Follansbee, 1986)

$$\bar{\sigma}\left(\bar{\varepsilon}_{pl}, \dot{\bar{\varepsilon}}_{pl}\right) = \sigma_y\left(\bar{\varepsilon}_{pl}\right)R\left(\dot{\bar{\varepsilon}}_{pl}\right) \quad (12)$$

where R is a stress ratio. Both shear failure and tensile failure were used to simulate the failure mechanisms of aluminium alloys (Vo et al., 2012).

Another constitutive relationship, namely the Cowper-Symonds relation, can be employed, if only the strain rate effect is considered. In this relation, strain rate is calculated for time duration from the start to the point when the strain is nearly constant from the equivalent plastic strain-time history (Cowper and Symonds, 1957), i.e.

$$\sigma_{dy} = \bar{\sigma}\left(1 + \left|\frac{\dot{\bar{\varepsilon}}_{pl}}{D}\right|^{1/n}\right) \quad (13)$$

where σ_{dy} is the dynamic yield stress, D and n are material constants.

3.2 Modelling fibre reinforced materials

Fibre reinforced composite laminates are often modelled as an anisotropic linear elastic material prior to the onset of damage, followed by damage evolution controlled either by fracture energy or element-based failure displacement. A damage mechanics approach may be suitable to predict composite damage in high strain rate loading conditions, such as blast. Kachanov (1958) and Rabotnov (1968) originally developed the damage mechanics approach, which was initially applied to composites by Frantziskonis (1988). It can be used to accurately determine the full range of deterioration in a composite material, from no damage to full damage with material disintegrated. Neto et al. (1998) and Krajcinovic (2000) gave a detailed review of issues in the area of damage mechanics and a historical overview. This approach has been used to predict different composite failure modes, such as matrix cracks, fibre fracture and delamination and even compression behaviour. Hashin's failure criteria (Hashin and Rotem, 1973; Hashin, 1980) are the popular ones for modelling onset damage of fibre reinforced composites, which are implemented in a 2D form in commercial codes such as ABAQUS and LS-DYNA. These criteria employs four damage initiation mechanisms, i.e. fibre tension, fibre compression, matrix tension and matrix compression. Other mode-dependent failure criteria were also developed by researchers (Lee, 1982; Chritensen, 1997; Mayes and Hansen, 2001).

In order to take material response through its thickness into account, Vo et al. (2012) implemented Hashin's 3D failure criteria into ABAQUS/Explicit through an user-defined subroutine, which are

Tensile fibre mode: $\sigma_{11} > 0$

$$\text{If}\left(\frac{\sigma_{11}}{X_{1t}}\right)^2 + \left(\frac{\sigma_{12}}{S_{12}}\right)^2 + \left(\frac{\sigma_{13}}{S_{13}}\right)^2 = 1,\ d_{ft} = 1 \quad (14)$$

Compressive fibre mode: $\sigma_{11} < 0$

$$\text{If}\frac{|\sigma_{11}|}{X_{1c}} = 1,\ d_{fc} = 1 \quad (15)$$

Tensile matrix mode: $\sigma_{22} + \sigma_{33} > 0$

$$\text{If}\frac{(\sigma_{22} + \sigma_{33})^2}{X_{2t}^2} + \frac{\sigma_{23}^2 - \sigma_{22}\sigma_{33}}{S_{23}^2} + \frac{\sigma_{12}^2 + \sigma_{13}^2}{S_{12}^2} = 1,\ d_{mt} = 1 \quad (16)$$

Compressive matrix mode: $\sigma_{22} + \sigma_{33} < 0$

$$\text{If}\left[\left(\frac{X_{2c}}{2S_{23}}\right)^2 - 1\right]\frac{(\sigma_{22} + \sigma_{33})}{X_{2c}} + \frac{(\sigma_{22} + \sigma_{33})^2}{4S_{23}^2} + + \frac{(\sigma_{23}^2 - \sigma_{22}\sigma_{33})}{S_{23}^2} + \frac{\sigma_{12}^2 + \sigma_{13}^2}{S_{12}^2} = 1,\ d_{mc} = 1 \quad (17)$$

In these failure criteria, damage variables d_{ft} and d_{fc} associate with fibre tension and compression failure modes respectively; and d_{mt} and d_{mc} associate with the corresponding failure modes in the matrix. Besides, lamina strength allowable values for tension and compression in the lamina principle material directions (fibre or 1-direction and matrix or 2-direction) as well as the in-plane shear strength allowable value are denoted by X_{1t}, X_{1c}, X_{2t}, X_{2c}, and S_{12}, respectively, and S_{13} and S_{13} are allowable values for the transverse shear strength. The degradation of stiffness matrix (**C**) (Tong et al., 2002) components due to fibre and matrix failure can be expressed as

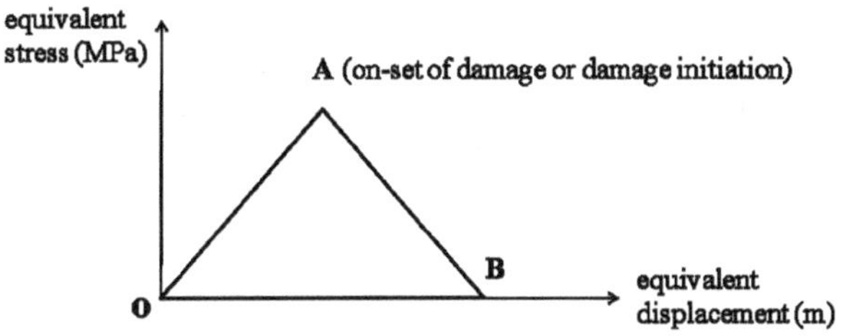

Figure 4. Energy dissipated due to damage.

$$
\begin{aligned}
C_{11} &= (1-d_f)C_{11}^0 \\
C_{ij} &= (1-d_f)(1-d_m)C_{ij}^0,\ i,j=1,2,3 \\
C_{ii} &= (1-d_f)(1-s_{mt}d_{mt})(1-s_{mc}d_{mc})C_{ii}^0,\ i=4,5,6
\end{aligned} \tag{18}
$$

where global fibre and matrix damage variables are also defined as:

$$
\begin{aligned}
d_f &= 1-(1-d_{ft})(1-d_{fc}) \\
d_m &= 1-(1-d_{mt})(1-d_{mc})
\end{aligned} \tag{19}
$$

They are either taken as 0 (no damage) and 1 (complete damage) in the code. The factors s_{mt} and s_{mc} in the definitions of the shear moduli are introduced to control the loss of shear stiffness caused by tensile and compressive failure of the matrix respectively.

Iannucci (2006) implemented a damage model into the explicit dynamic FE code LLNL DYNA3D (Hallquist and Whirley, 1989) to predict a woven composite on a layer-by-layer basis (2D), with damage parameters as:

- Fibre-fracture in the local warp fibre bundles for each ply, d_1
- Fibre-fracture in the local weft fibre bundles for each ply, d_2
- Fibre-matrix deterioration due to in-plane shear for each ply, d_3

The damage variables are in the range from 0 to 1, with the former representing an undamaged material and the latter a complete failure. The damage terms in the stiffness degradation matrix are as follows.

$$
\frac{1}{Z}\begin{bmatrix} (1-d_1)E_{11}^0 & (1-d_2)(1-d_1)\nu_{12}^0E_{22}^0 & 0 \\ (1-d_2)(1-d_1)\nu_{21}^0E_{11}^0 & (1-d_1)E_{11}^0 & 0 \\ 0 & 0 & ZG_{12}^0(1-d_1) \end{bmatrix} \tag{20}
$$

where $Z=1-(1-d_1)(1-d_2)\nu_{12}^0\nu_{21}^0>0$. The superscript zero represents material properties in the virgin state.

The most simple damage evolution is modelled by the negative slope of the equivalent stress-displacement relation after damage initiation is achieved, as shown in Fig. 4. Fracture energies for fibre tension, fibre compression, matrix tension and matrix compression failure modes need to be specified to indicate energies dissipated during damage. Nahas (1986) carried out progressive failure analysis based on the failure predictions of a modified maximum strain criterion. The degradation function defined in Eq. (21) is displayed in Fig. 5.

$$
\sigma = \sigma_{\max} e^{(-a(\varepsilon-\varepsilon_{\max})/n\varepsilon_{\max})} \tag{21}
$$

Here, a and n are empirical constants. Iannucci (2006) proposed a 2D damage evolution law for all damage modes in the local coordinate system, which is given by

$$
\dot{d}_i = \alpha_i\left(\left(\frac{\bar{\sigma}_{jk}}{S_{jk}}\right)^2-1\right) \tag{22}
$$

where d_i is the damage parameters for mode i (i = 1,2,3), α_i and S_{jk} (j,k = 11, 22, 12) are material constants in this simple form. For initiation of nucleation, effective stress $\bar{\sigma}_{jk}$ meeds to be greater than the in-plane lamina strength S_{jk}. A similar form to Eq. (22) was also proposed by Nemes (1996). A thorough review on degradation models for progressive failure analysis of fibre reinforced polymer composites is given by Garnich and Akula (Garnich and Akula, 2009).

3.3 Modelling cellular solids

Metal foams, PVC foams and honeycombs are cellular solids, which can absorb considerable energy through plastic dissipation in compression. Their cellular microstructures make them undergo large deformations at nearly constant nominal stress (Gibson and Ashby, 1997; Hanssen et al., 2002). As the result, they are usually used as core materials for sandwich structures in order to absorb impact and blast shock energy. Constitutive relations are given in terms of the effective stress (Lemaitre, 1996) as

$$
\bar{\sigma} = Y^0 + Q_1(1-e^{-c_1 r}) + Q_2(1-e^{-c_2 r}) \tag{23}
$$

where Y^0 is the uniaxial yield stress, r_1 and r_2 are the damage plastic strain accumulated, the rest are the hardening parameters. Based on a self-similar yield surface model, Deshpande and Fleck (2000) developed a constitutive model to describe metal foam undergoing continuous crushing, which is given by

$$
\sigma_y^2 \equiv \frac{1}{\left[1+\left(\frac{\alpha}{3}\right)^2\right]}\left[\bar{\sigma}^2+\alpha^2\sigma_m^2\right] \tag{24}
$$

Here, α defines the shape of the yield surface, σ_m is the mean stress, and σ_y is the yield strength of the foam in uniaxial tension or compression. This constitutive model was also used to model truss core (Xue and Hutchinson, 2003) and aluminium honeycomb core (Zhu et al., 2009) and PVC foam core (Hassan et al., 2012, Langdon et al., 2013) subjected to blast loading.

3.4 Rate dependence

Fibre reinforced composites are essentially rate-dependent materials when they are subjected to high strain rates,

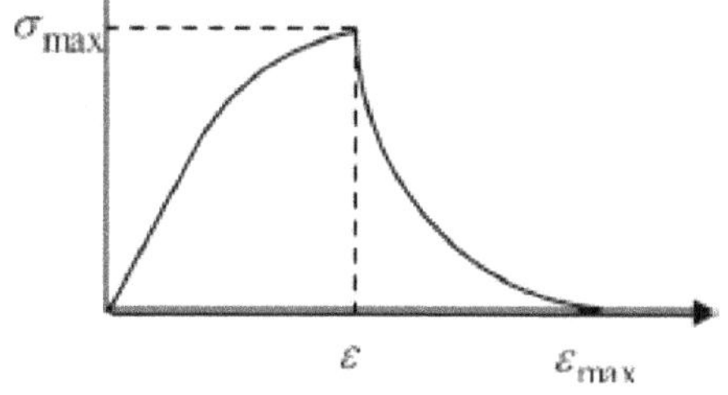

Figure 5. Nahas unloading model.

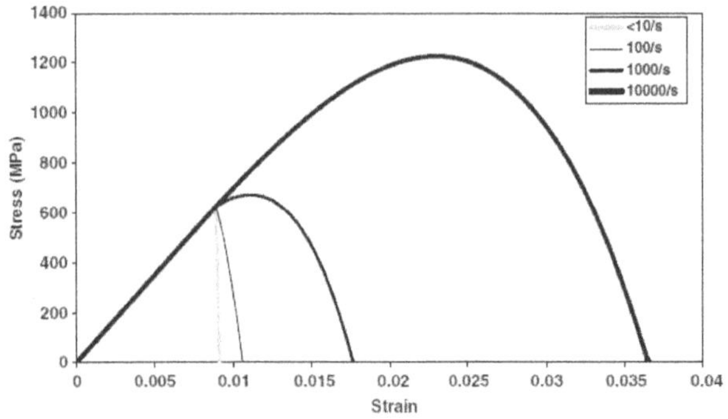

Figure 6. Strain-rate dependence for tensile loadings.

such as under blast loading (Harding, 1989; Li and Lambros, 2000). General conclusions indicate that an increase in strain-rate induces an increase in tensile failure stress and strain, however, the initial modulus appears to be unaffected. Fig. 6 shows the strain-rate response with material constants set as α_i = 1000/s and S_{jk} = 610MPa (Eq. (22)). The curves are generated via a single element subject to varying strain-rates.

Based on the work by Yen and Caiazzo (2001), Yen (2002) and Daniel et al. (2011), Vo et al. (2012) considered strain rate effects on material propteris and strengths as follows:

$$\{E_{RT}\} = \{E_0\}\left(1 + C_1 \ln\frac{\dot{\varepsilon}}{\dot{\varepsilon}_0}\right) \quad (25)$$

$$\{S_{RT}\} = \{S_0\}\left(1 + C_2 \ln\frac{\dot{\varepsilon}}{\dot{\varepsilon}_0}\right) \quad (26)$$

Here, C_1 and C_2 is the strain rate constants, $\dot{\varepsilon}_0$ is reference strain rate, $\{E_0\}$ are elastic moduli of $\{E_{RT}\}$ at the reference strain rate and $\{S_0\}$ are strength values of $\{S_{RT}\}$ at the reference strain rate.

4. Modelling work undertaken

Extensive numerical modelling has been undertaken on structural responses of composite laminates and sandwich structures subjected to blast. The work is mainly carried out by using commercial codes with built-in constitutive models, failure criteria and damage evolution or by developing in-house programmes. Karagiozova et al. (2010) using the ABAQUS/Explit code simulated the response of fibre metal laminates (FMLs) subjected to low impulsive localised blast loading. Blast responses of GLARE fuselage (Kotzakolios et al., 2011) and panels (Soutis et al., 2011) were also simulated using LS-DYNA. A nonlinear transient analysis of FMLs under blast loads was carried out using the mixed FEM by Aksoylar et al. (2012). Vo et al. (2012; 2013) developed a vectorized user material subroutine (VUMAT) and implemented it in ABAQUS/ Explicit to model blast behaviour of woven glass-fibre in a polypropylene (GFPP) composite inside FMLs, with strain rate effects considered. Sitnikova et al. (2013) developed an instant failure model to simulate perforation failure of GFPP layers inside FMLs subjected to high implusive blast loading. A number of dynamic failure scenarios were captured, such as petalling, large tensile tearing and multiple debonding between the aluminium and GFPP layers. Fig. 7 shows a typical comparison between the predicted displacement contour plots with the experimental data.

Modelling work using commercial code ABAQUS/ Explicit was undertaken on blast responses of metal foam based sandwich structures (Xue and Htchinson, 2003; Qiu et al., 2003; Qiu et al., 2004; Radford et al., 2006), metal honeycomb based sandwich structures (Mori et al., 2007; Wei et al., 2007; Theobald and Nurick, 2010), as well as PVC foam based sandwich structures (Hassan et al., 2012; Longdon et al., 2013). Fig. 8 shows some simulated and experimental failure modes of PVC foam based sandwich panels with aluminium alloy skins. LS-DYNA was also used to model blast behaviour of metal foam based sandwich panels (Zhu et al., 2009; Jing et al., 2013), skull sandwich structures (Gu et al., 2012) and composite laminates (Wei et al., 2006; LeBlane and Shukla, 2011). 3-D MOSAIC analysis approach was used to investigate shock loading response of sandwich panels with 3-D woven composite skins and stitched foam core (Tekalur et al., 2009).

Some researchers developed in-house numerical programmes to simulate responses of various composite structures under explosive loads. Librescu et al. (2004;

Figure 7. Comparison of the predicted displacement contour plots with the experimental data.

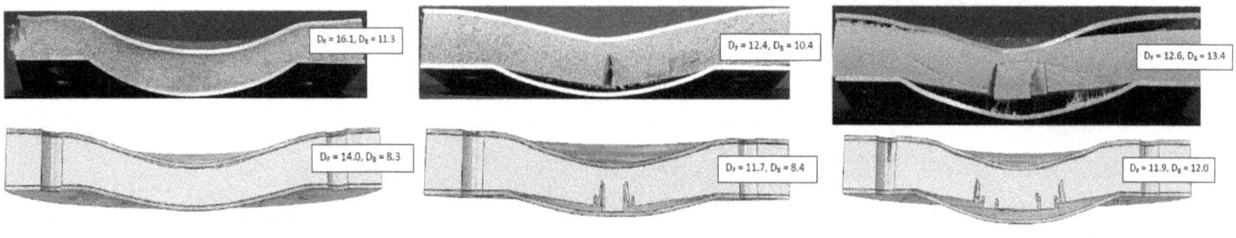

ρ = 60 kg/m³, I = 8.3 Ns ρ = 130 kg/m³, I = 8.3 Ns ρ = 200 kg/m³, I = 13.1 Ns

Figure 9. Comparisons of failure modes obtained from tests (top) and FE simulations (bottom).

2008) studied dynamic behaviour of sandwich panels and active aeroelastic control of aircraft composite wings to blast loading, with governing equation systems in the von Kármán sense. Sandwich plates/panels subjected to blast loadings were also modelled numerically by the basic equations of the dynamic theory of advanced curved panels (Hause and Librescu, 2007) and laminated composite rectangular plates (Kazanci, 2011). In addition, in-house finite element codes were developed to model blast response of fibre reinforced composites (Batra and Hassan, 2007; 2008) and sandwich plates (Nayak et al., 2006; Andrews and Moussa, 2009) to underwater explosive loads and air blast loads.

5. Summary

Numerical modelling work of fibre reinforced composite laminates and various sandwich structures subjected to blast loadings has been reviewed. Before carry out modelling of structural response, explosive loading has to be simplified to reflect the most important characteristics. Several approaches commonly adopted, either in chart or equation forms, are discussed. A majority of the estimations on the effective duration of blast loading is 10s of microseconds (usually less than 40 μs), however some are assumed in minisecond (ms). As there is no a precise measurement of such the duration, even taking it as 10 or 20 or 30 μs, there will be a big difference on calculations of the impulsive pressure. This affects numerical modelling greatly. Therefore, in order to simulate blast response accurately, the reasonable estimation of the blast duration is crucial.

Modelling of metallic constituent materials is relatively simple, since commonly used constitutive models and failure criteria are available in commercial FE codes. Johnson-Cook rate-dependent plasticity with damage (degradation) measures is a good model, which is capable of capturing various failure modes of metals. However, accurate material parameters, covering rate-dependent plastic hardening, temperature-dependent behaviour as well as damage constants, are essential. In addition, modelling of cellular structures, such as metal foams, truss cores, honeycombs and polymer foams, seems well addressed. The crushable foam model has been approved by a large number of researches as an effective approach to simulate the corresponding blast response.

In comparison, modelling blast behaviour of fibre reinforced composites still remains a challenge task, especially involving simulation of perforation failure. Most of the published work deals with the blast response in relation to relatively low impulsive pressures, which do not cause fantastic failure of fibre reinforced composites. Yet, mainly only 2-D failure criteria and damage evolution are available in commercial FE codes, which do not take the structural response through the thickness into account. The possible reason is that the failure mechanisms of fibre debonding and breaking under an extremely high strain rate are not well captured in the existing programmes. It seems fine to simulate the response up to the onset of failure, also delamination between layers by cohesive elements. Therefore, it is necessary to develop and implement 3-D failure criteria and damage evolution laws with appropriate description of high strain rate dependence to simulate progressive failure of fibres with numerical stability (convergent rate).

References

Aksoylar, C., Ömercikoğlu, A., Mecitoğlu, Z. and Omurtag, M.H. (2012). Nonlinear transient analysis of FGM and FML plates under blast loads by experimental and mixed FE methods, *Composite Structures*, Vol. 94, pp. 731-744.

Andrews, E.W. and Moussa, N.A. (2009). Failure mode maps for composite sandwich panels subjected to air blast loading, *Int. J. Impact Eng.*, Vol. 36, pp. 418–425.

Batra , R.C. and Hassan, N.M. (2007). Response of fiber reinforced composites to underwater explosive loads, *Composites: Part B, Vol. 38*, pp. 448–468.

Batra , R.C. and Hassan, N.M. (2008). Blast resistance of unidirectional fiber reinforced composites, *Composites: Part B*, Vol. 39, pp. 513–536.

Bonorchis, D. and Nurick, G.N. (2009). The influence of boundary conditions on the loading of rectangular plates subjected to localised blast loading - Importance in numerical simulations, *International Journal of Impact Engineering*, Vol. 36, No. 1, pp. 40 - 52.

Brode, H.L. (1955). *Numerical Solution of Spherical Blast Waves*, American Institute of Physics, New York (Journal of Applied Physics).

Chirica, I., Boazu, D. and Beznea, E.F. (2012). Response of ship hull laminated plates to close proximity blast loads, *Computational Materials Science*, Vol. 52, pp. 197-203.

Christensen, R. M. (1997). Stress Based Yield/Failure Criteria for Fiber Composites, *Int. J. Solids Struct.*, Vol. 34, No. 5, pp. 529–543.

Cowper, G.R. and Symonds, P.S. (1957). *Strain Hardening and Strain Rate Effect in the Impact Loading of Cantilever Beams*, Brown University, Division of Applied Mathematics, Report No. 28.

Daniel, I.M., Werner, B.T. and Fenner, J.S. (2011). Strain-rate-dependent failure criteria for composites, *Composites Science and Technology*, Vol. 71, No. 3, pp. 357 -- 364.

Deshpande, V.S. and Fleck, N.A. (2000). Multi-axial yield behaviour of polymer foams, *Journal of Mechanics and Physics of Solids*, Vol. 48, pp. 1859-1866.

Frantziskonis, S. (1988). Distributed damage in composites, theory and verification, *Composite Struct*, Vol. 10, No. 2, pp. 165–184.

Garnich, M.R. and Akula, V.M.K. (2009). Review of degradation models for progressive failure analysis of fiber reinforced polymer composites, *Applied Mechanics Review*, Vol. 62, pp. 010801-1 – 33.

Gibson, L.J. and Ashby, M.F. (1997). *Cellular Solids: Structure and Properties* (2nd edn.), Cambridge: Cambridge University Press.

Gu, L., Chafi, M.S., Ganpule, S. and Chandra, N. (2012). The influence of heterogeneous meninges on the brain mechanics under primary blast loading, *Composites: Part B*, Vol. 43, pp. 3160-3166.

Hallquist, J.O. and Whirley, R.G. (1989). *DYNA3D User Manual, Nonlinear Dynamic Analysis in Three Dimensions*, University of California, Lawrence Livermore National Laboratory, Report UCID-19592, Rev. 5.

Harding, J. (1989). Impact damage in composite materials, *Sci Eng Composite Mater*, Vol. 1, pp. 41–68.

Hashin, Z., and Rotem, A. (1973). A fatigue criterion for fiber-reinforced materials, *Journal of Composite Materials*, Vol. 7, pp. 448 –464.

Hashin, Z. (1980). Failure criteria for unidirectional fiber composites, *Journal of Applied Mechanics*, Vol. 47, pp. 329–334.

Hassan, M.Z., Guan, Z.W., Cantwell, W.J., Langdon, G.S. and Nurick, G.N. (2012). The influence of core density on the blast resistance of foam-based sandwich structures, *International Journal of Impact Engineering*, Vol. 50, pp 9-16.

Hause, T. and Librescu, L. (2007). Dynamic response of doubly-curved anisotropic sandwich panels impacted by blast loadings, *International Journal of Solids and Structures*, Vol. 44, pp. 6678–6700.

Hyde, D. (1992). *ConWep, Conventional Weapons Effects Program*, Vicksburg (MS): US Army Engineer Waterways Experiment Station.

Iannucci, L. (2006). Progressive failure modelling of woven carbon composite under impact, *International Journal of Impact Engineering*, Vol. 32, pp. 1013–1043

Jing, L., Wang, Z. and Zhao, L. (2013). Dynamic response of cylindrical sandwich shells with metallic foam cores under blast loading—Numerical simulations, *Composite Structures*, Vol. 99, pp. 213-223.

Johnson, G.R. and Cook, W.H. (1983). A constitutive model and data for metals subjected to large strains, high strain rates and high temperatures, *Proceedings of the 7th International Symposium on Ballistics*, the Hague, the Netherlands.

Johnson, G.R. and Cook, W.H. (1985). Fracture characteristics of three metals subjected to various strains, strain rates, temperatures and pressures, *Engineering Fracture Mechanics*, Vol. 21, No. 1, pp. 31-48.

Kachanov, L.M. (1958). Time of rupture process under creep conditions, *Izy Akad Nank U.S.S.R.* Otd Tech Nauk, Vol. 8, pp. 26-31.

Karagiozova, D., Langdon, G.S., Nurick, G.N. and Yuen, S.C.K. (2010). Simulation of the response of fibre-metal laminates to localised blast loading, *International Journal of Impact Engineering*, Vol. 37, No. 6, pp. 766 – 782.

Kazanci, Z. (2011). Dynamic response of composite sandwich plates subjected to time-dependent pressure pulses, *International Journal Non-linear Mechanics*, Vol. 46, pp. 807-817.

Kinney, G.F. (1962). *Explosive Shocks in Air*, New York: The Macmilian Co.

Kinney, G.F., and Graham, K.J. (1985) *Explosive Shocks in Air*, 2nd edn., New York: Springer-Verlag.

Kotzakolios, T., Vlachos, D.E. and Kostopoulos, V. (2011). Blast response of metal composite laminate fuselage structures using finite element modelling, *Composite Structures*, Vol. 93, pp. 665-681.

Krajcinovic, D. (2000). Damage mechanics: accomplishments, trends and needs, *International Journal Solids Structures*, Vol. 37, No. 1–2, pp. 267–77.

Langdon, G.S., Nurick, G.N., Lemanski, S.L., Simmons, M.C., Cantwell, W.J. and Schleyer, G.K. (2007). Failure characterisation of blast-loaded fibre-metal laminate panels based on aluminium and glass-fibre reinforced polypropylene, *Composites Science and Technology*, Vol. 67, No. 7-8, pp. 1385-1405.

Langdon, G.S., Karagiozova, D., von Klemperer, C.J., Nurick, G.N., Ozinsky, A. and Pickering, E.G. (2013). The air-blast response of sandwich panels with composite face sheets and polymer foam cores: Experiments and predictions, *International Journal of Impact Engineering*, Vol. 54, pp 64-82.

LeBlanc, J. and Shukla, A. (2011). Dynamic response of curved composite panels to underwater explosive loading: Experimental and computational comparisons, *Composite Structures*, Vol. 93, pp. 3072 -3081.

Lee, J. D. (1982). Three-Dimensional Finite Element Analysis of Damage Accumulation in Composite Laminates, *Proceedings of the Second USAUSSR Symposium*, Bethlehem, PA, Mar. 9–12, pp. 291–306.

Lemaitre, J. (1996). A course on damage mechanics, 2nd ed. Berlin: Springer, 1996, ISBN 3-540-60980-6.

Li, Z. and Lambros, J. (2000). Dynamic thermomechanical behaviour of fiber reinforced composites, *Composites Part A*, Vol. 31, pp. 537–47.

Librescu, L., Oh, S.Y. and Hoheb, J. (2004). Linear and non-linear dynamic response of sandwich panels to blast loading, *Composites: Part B*, Vol. 35, pp. 673–683.

Librescu, L., Na, S., Qin, Z. and Lee, B. (2008). Active aeroelastic control of aircraft composite wings impacted by explosive blasts, *Journal of Sound and Vibration*, Vol. 318, pp. 74–92.

Marzocca, P., Librescu, L., Chiocchia, G. (2001). Aeroelastic response of 2D lifting surfaces to gust and arbitrary explosive loading signature, *International Journal Impact Engineering*, Vol. 25, No. 1, pp. 67-85.

Mori, L.F., Lee, S., Xue, Z.Y., Vaziri, A., Queheillalt, D.T., Dharmasena, K.P., Wadley, H.N.G., Hutchinson, J.W. and Espinosa, H.D. (2007). Deformation and fracture modes of sandwich structures subjected to underwater impulsive loads, *Journal of Mechanics of Materials and Structures*, Vol. 2, No. 10, pp. 1981-2006.

Mayes, J. S. and Hansen, A. C. (2001). A Multicontinuum Failure Analysis of Composite Structural Laminates, *Mech. Compos. Mater. Struct.*, Vol. 8, No. 4, pp. 249–262.

Nahas, M. N. (1986). Yield and Ultimate Strengths of Fibre Composite Laminates, *Compos. Struct.*, Vol. 6, No. 2, pp. 283–294.

Nayak, A.K., Shenoi, R.A. and Moy, S.S.J. (2006). Transient response of initially stressed composite sandwich plates, *Finite Elements in Analysis and Design*, Vol. 42, pp. 821-836.

Nemes, J.A. (1996). Use of a rate-dependent continuum damage model to describe strain-softening in laminated Composites, *Comput Struct*, Speciel E., Vol. 58, pp. 1083–92.

Neto, E., Peric, D. and Owen, D.R.J. (1998). Continuum modelling and numerical simulation of material damage at finite strains, *Arch Comput Methods in Eng*, Vol. 5, No. 4, pp. 311–84.

Qiu, X., Deshpande, V.S. and Fleck, N.A. (2003). Finite element analysis of the dynamic response of clamped sandwich beams subject to shock loading, *Eur. J. Mech. A/Solid*, Vol. 22, pp. 801-814.

Qiu, X., Deshpande, V.S., Fleck, N.A. (2004). Dynamic response of a clamped circular sandwich plate subject to shock loading, *Journal of Applied Mechanics, ASME*, Vol. 71, pp.637-645.

Rabotnov, Y.N. (1968). Creep rupture, *Proceedings of the XII International Congress Applied Mechanics*, Springer: Stanford.

Radford, D.D., Mcshane, G.J., Deshpande, V.S. and Fleck, N.A. (2006). The response of clamped sandwich plates with metallic foam cores to simulated blast loading, *Int. J. Solids and Structures*, Vol. 43, pp. 2243-2259.

Rajamani, A. and Prahakaran, R. (1980). Response of composite plates to blast loading, *Experimental Mechanics*, Vol. 20, No. 7, pp. 245-250.

Sitnikova, E., Guan, Z.W., Cantwell, W.J. and Schleyer, G.K. (2013). Modelling of perforation failure in fibre metal laminates subjected to high impulsive blast loading, submitted to *Journal of Mechanics and Physics of Solids*.

Soutis, C., Mohamed, G. and Hodzic, A. (2011). Modelling the structural response of GLARE panels to blast load, *Composite Structures*, Vol. 94, pp. 267-276.

Stout, M.G. and Pollansbee, P.S. (1986). Strain rate sensitivity, strain hardening, and yield behaviour of 304L stainless steel, Transactions of ASME: *Journal of Engineering Materials Technology*, Vol. 108, pp. 344-353.

Tekalur, S.A., Bogdanovich, A.E. and Shukla, A. (2009). Shock loading response of sandwich panels with 3-D woven E-glass composite skins and stitched foam core, *Composite Sci. and Tech.*, Vol. 69, pp.736-753.

Theobald, M.D. and Nurick, G.N. (2010). Experimental and numerical analysis of tube-core claddings under blast loads, *Int. J. Impact Eng.*, Vol. 37, pp. 333-348.

Thiagarajan, G., Hsia, K.J. and Walters, W. (2003). Blast simulation of armor plate behavior using the virtual internal bond model, 16th *ASCE Engineering Mechanics Conference*, July 16-18, University of Washington, Seattle.

Tong, L., Mouritz, A.P. and Bannister, M.K. (2002). *3D Fibre Reinforced Polymer Composites*, London: Elsevier.

Vo, T.P., Guan, Z.W., Cantwell, W.J. and Schleyer, G.K. (2012). Low-impulse blast behaviour of fibre-metal laminates, *Composite Structures*, Vol. 94, pp. 954-965.

Vo, T.P., Guan, Z.W., Cantwell, W.J. and Schleyer, G.K. (2013). Modelling of the low-impulse blast behaviour of fibre–metal laminates based on different aluminium alloys, *Composites: Part B*, Vol. 44, pp. 141-151.

Wei, Z., He, M.Y. and Evans, A.G. (2007). Application of a Dynamic Constitutive Law to Multilayer Metallic Sandwich Panels Subject to Impulsive Loads, *Journal of Applied Mechanics, ASME*, Vol. 74, pp.636-644.

Wei, Z., Shetty, M.S. and Dharani, L.R. (2006). Stress characteristics of a laminated architectural glazing subjected to blast loading, *Computers and Structures*, Vol. 84, pp. 699–707.

Xue, Z. and Hutchinson, J.W. (2003). Preliminary assessment of sandwich plates subject to blast, *Int. J. Mech. Sci.*, Vol. 45, pp. 687-705.

Yen, C.F. and Caiazzo, A. (2001). *Innovative Processing of Multifunctional Composite Armor for Ground Vehicles*, ARL-CR-484, US Army Research Laboratory, Aberdeen Proving Ground, MD.

Yen, C.F. (2002). Ballistic impact modeling of composite materials, in: *Proceedings of the 7th International LS-DYNA Users Conference*, Vol. 6, pp. 15-23.

Zhu, F., Zhao, L., Lu, G. and Gad, E. (2009). A numerical simulation of the blast impact of square metallic sandwich panels, International *Journal of Impact Engineering*, Vol. 36, pp. 687–699.

Theoretical Rationale and Experimental Verification of the Method for Identification of Thermal Conductivity in High-temperature Anisotropic Composite Materials

Pavel Prosuntsov, Sergey Reznik

Rocket and Spacecraft Composite Structures Department, Faculty of Special Machinery, Bauman Moscow State Technical University, 5 2nd Baumanskaya Street, Moscow, 105005, Russia

Abstract: In order to control technological processes and to design constructions out of composite materials (CM), one has to possess the data on the thermophysical properties (TPP) of the materials as well as their structural charecteristics. The current paper presents a method for complex identification of the thermal conductivity coefficient for anisotropic composite materials in the conditions of variable heating received by the experimental samples. The method under consideration is based on the solution of a two dimensional non-linear inverse problem of transient heat conductivity. Different schemes of thermal tests were analysed with the use of a temperature measurement planning method based on the theory of information matrices. Two types of sample heating were considered: direct-resistance heating and electrical source heating. Water cooler heat exchangers and partial shielding of the surface were used in order to ensure temperature gradients both in the longitudinal and transversal directions of the sample. There has been run a comparison between various schemes. There has also been provided a rationalization for the choice of thermal test conditions. The current research provides a characteristic of an automated thermophysical complex which allows heating of the samples with the speeds of up to 100 K/s from room temperature to 1700 K. Experimental data processing results and their analysis are presented.

Key Words: Composite materials, Heat conductivity coefficient, Anisotropic characteristics.

1. Introduction

Heat resistant carbonaceous composite materials (CM) are widely used in aerospace structures (Reznik et al., 2002). They are characterized by high strength-weight and stiffness-weight ratio as well as a substantial anisotropy of resources. Still with no protective coating they get oxidized in the air at high temperatures. In order to ensure the control over the technological processes of material manufacture and to design constructions out of them it is necessary to possess data on the thermophysical properties (TPP) in connection with their thermal life history and structural characteristics. Methods of modelling and parametrical identification of thermophysical processes in material samples and construction elements are a powerful instrument of conducting multiscale research like this (Alifanov et al., 2001).

Among acute problems one can name complex identification of temperature dependences for heat conductivity coefficients, specific heat and radiating capacity of CM in a wide temperature range by methods which are highly productive and accurate at the same time. There exists a variety of ways used for studying the TPPs of materials (Osipova, 1979, Fillipov, 1984, Platunov et al., 1986). In the classical method of the plate aimed at identifying thermal conductivity coefficients, measurements are taken at plane samples in one of the directions while establishing fixed thermal conditions. A one-dimensional temperature field is formed with the help of a highly accurate means of temperature control, radiation shields, protective heat insulation of high-porous materials and vacuuming the measurement zone. In long-term experiments certain difficulties arise. They are related to the necessity of maintaining a one-dimensional fixed temperature field in the high temperature area. Protective heat insulation of side surfaces loses its efficiency, shrinks and gets into the thermochemical contact with the sample.

Electroconductive and relatively highly heat-conducting carbonaceous CMs can be heated to high temperatures either by direct-resistance heating or with the help of concentrated energy sources – solar installations, arc and halogen lamps with reflectors, plasmatrons (Reznik et al., 2002). The problem of using traditional approaches to processing experimental data is related to the fact that in the conditions of intensive heating the temperature state of CM samples rapidly changes both in time and in space coordinates. In addition to that the placement of temperature sensors inside the experimental samples is undesirable or impossible due to a number of reasons.

2. Mathematical modelling of heat transfer in a sample

We are going to consider the process of heat transfer in a plane sample of anisotropic composite material. It is presumed that experimental conditions ensure two-dimensional heat transfer in the sample. Thermophysical properties of the sample material (thermal conductivity coefficients in two directions λ_x and λ_y and volumetric heat capacity C) depend on the temperature, so do the optical properties of its surface (absorptive capacity A and emissivity ε). The experiment studies heat transfer in a rectangular two-dimensional area Ω with boundary $\partial\Omega$, while at a part of the area boundary $\partial\Omega_1$ first-type boundary conditions are set and at a part of the boundary $\partial\Omega_2$ second-type boundary conditions are set. The sample can be heated by the influence of external radiant flux q_w or by ohmic heat while running electric current through the sample, in the latter case there takes place a heat release with a capacity of q_V in the volume of the sample. On the surfaces of the sample in touch with the surrounding gaseous medium which has the temperature of T_f there takes place convective heat transfer with the heat transfer coefficient of α_f.

The following mathematical model can be formed for

the abovementioned heat transfer process:

$$C(T)\frac{\partial T(x,y,\tau)}{\partial \tau}=\frac{\partial}{\partial x}\left(\lambda_x(T)\frac{\partial T(x,y,\tau)}{\partial x}\right)+$$
$$+\frac{\partial}{\partial y}\left(\lambda_y(T)\frac{\partial T(x,y,\tau)}{\partial y}\right)+q_V(x,y,\tau), \tag{1}$$

$$(x,y)\in\Omega \quad \tau\in\left]0,\tau_f\right]$$

$$\tau=0 \; T(x,y,0)=T_0(x,y); \tag{2}$$

$$\partial\Omega_1\cap\partial\Omega \; T(\partial\Omega_1,\tau)=T_w(\partial\Omega_1,\tau); \tag{3}$$

$$\partial\Omega_2\cap\partial\Omega \; -\lambda(T)\frac{\partial T(\partial\Omega_2,\tau)}{\partial n}=$$
$$=A(T)q_w(\tau)-\varepsilon(T)\sigma_0\left(T^4(\partial\Omega_2,\tau)-T_f^4\right)- \tag{4}$$
$$-\alpha_f(T)\left(T(\partial\Omega_2,\tau)-T_f\right)$$

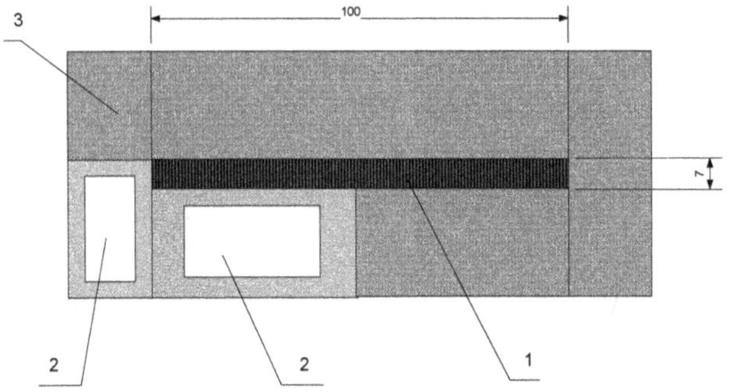

Figure 1. 1 – sample; 2 – water cooler; 3 – heat insulation. Conditions of the experiment: direct resistance heating at the capacity of auto thermal heat supplies of 20 MW/m^3; water cooler temperature 300 K; heating duration 300 s.

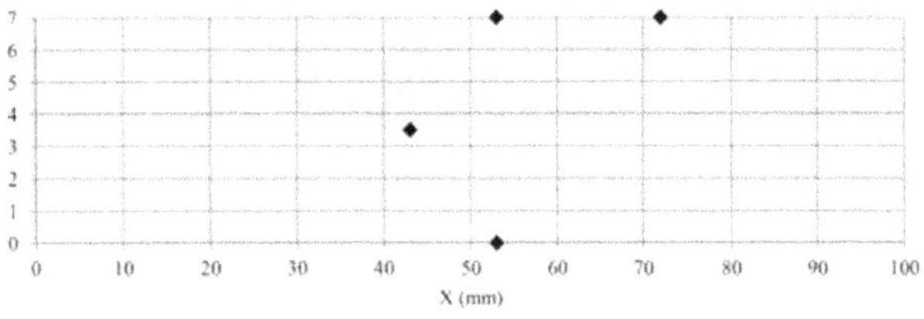

Figure 2. Coordinates of temperature sensors obtained as a result of solving experiment planning problem.

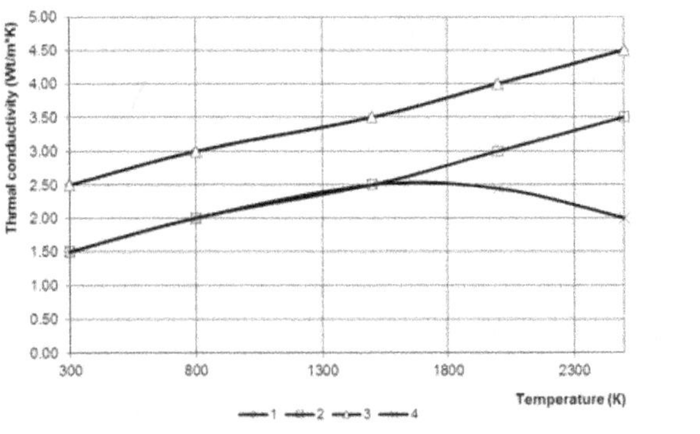

Figure 3. Solution results of the heat conductivity inverse problem for defining the temperature dependencies of thermal conductivity coefficient for the anisotropic material. 1 – dependence $\lambda_x(T)$ of the model material; 2 – dependence $\lambda_y(T)$ of the model material; 3,4 – dependences $\lambda_x(T)$ and $\lambda_y(T)$, obtained from the inverse problem solution.

3. Parametrical identification of thermophysical properties for anisotropic material

Let during the heating of the sample temperature measurements be taken in points. It is required to use these measurements to determine the temperature dependencies of heat conductivity coefficients in two directions.

In order to solve the inverse problem we are going to use the extreme statement and iterative regularization principle (Alifanov et al., 2001). I.e. we will construct the discrepancy functional discrepancy of the experimental and estimated temperature values, and we will define the temperature dependences of thermal conductivity coefficients for the anisotropic material which afford a minimum to this functional:

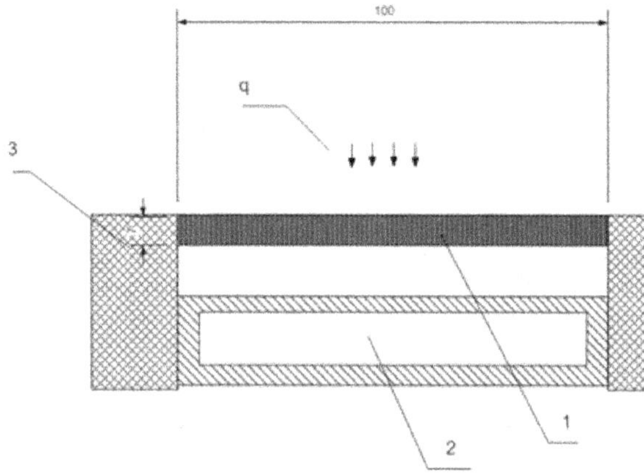

Figure 4. Scheme of the experiment. 1 – sample; 2 – water cooler; 3 – heat insulation. Conditions of the experiment: radiative heat flux heating with the density of 3 MW/m^2; water cooler temperature 300 K; heating duration 300 s.

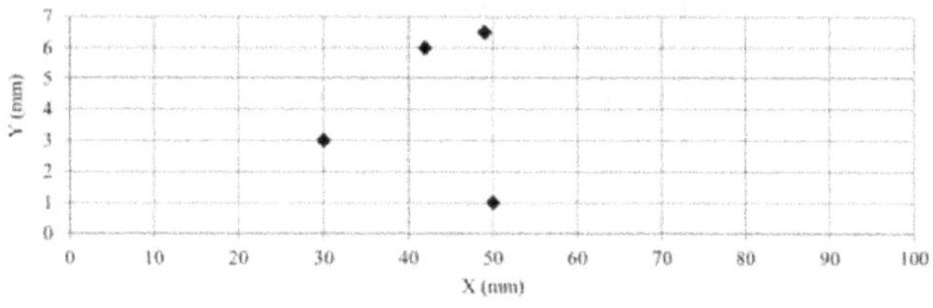

Figure 5. Coordinates of temperature sensors obtained as a result of solving experiment planning problem.

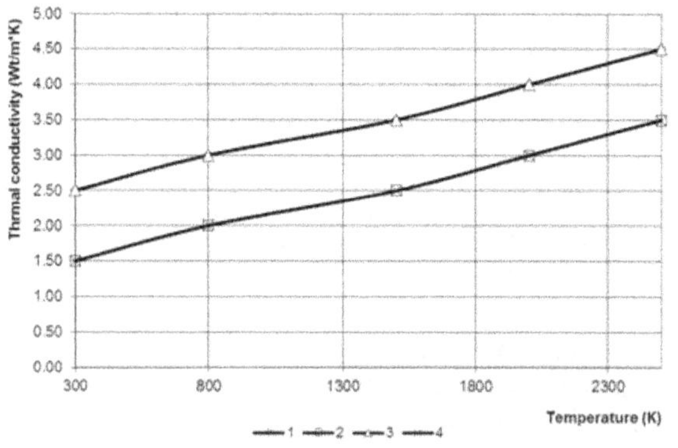

Figure 6. Solution results of the heat conductivity inverse problem for defining the temperature dependencies of thermal conductivity coefficient for the anisotropic material. 1 – dependence $\lambda_x(T)$ of the model material; 2 – dependence $\lambda_y(T)$ of the model material; 3,4 – dependences $\lambda_x(T)$ and $\lambda_y(T)$, obtained from the inverse problem solution.

$$S(\vec{u}) = \frac{1}{2}\int_0^{\tau_f}\sum_{n=1}^{N_t}\left(T\left(x_n^t, y_n^t, \tau\right) - T^e\left(x_n^t, y_n^t, \tau\right)\right)^2 d\tau; \qquad (5)$$

$$\vec{u} = \left\{\lambda_y(T), \lambda_x(T)\right\}$$

where τ_m stands for the duration of the experiment, t - for index indicating a temperature sensor, T^e - for experimentally measured temperature values.

Discrepancy method is used to stop the iteration process of functional minimization:

$$S(\vec{u}) = \min; \quad S \geq \Delta^2 \qquad (6)$$

where: Δ stands for the error of temperature measurement.

We will parameterize the sought temperature dependences of the thermal conductivity coefficient:

$$\lambda_x(T) = \sum_{i=1}^{K_{\lambda x}} \lambda_{xi}\chi_{1i}(T)$$

$$\lambda_y(T) = \sum_{j=1}^{K_{\lambda y}} \lambda_{yj}\chi_{2j}(T) \qquad (7)$$

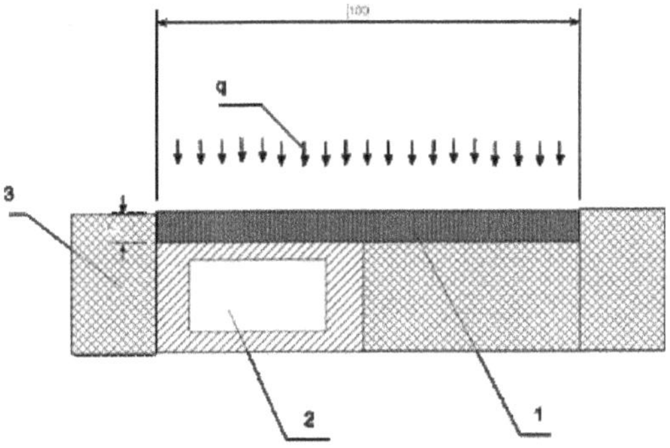

Figure 7. Scheme of the experiment. 1 – sample; 2 – water cooler; 3 – heat insulation. Conditions of the experiment: radiative heat flux heating with the density of 3 MW/m^2; water cooler temperature 300 K; heating duration 300 s.

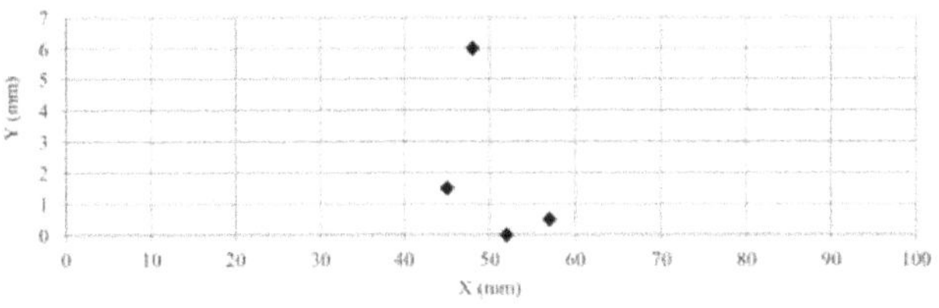

Figure 8. Coordinates of temperature sensors obtained as a result of solving experiment planning problem.

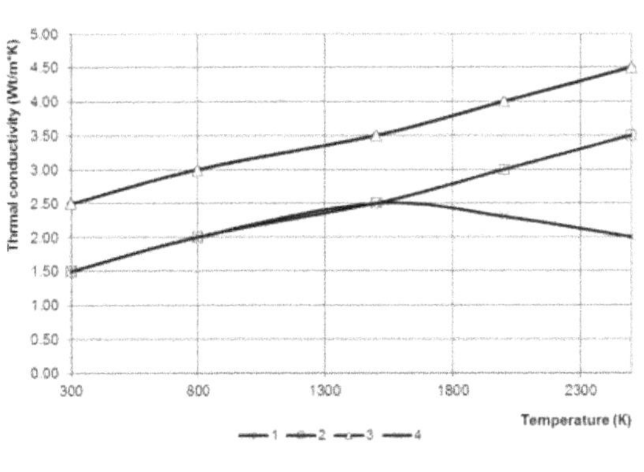

Figure 9. Solution results of the heat conductivity inverse problem for defining the temperature dependencies of heat conductivity coefficient for the anisotropic material. 1 – dependence $\lambda_x(T)$ of the model material; 2 – dependence $\lambda_y(T)$ of the model material; 3,4 – dependences $\lambda_x(T)$ and $\lambda_y(T)$, obtained from the inverse problem solution.

where: $\chi_{1j}(T)$, $\chi_{2j}(T)$ stand for basis functions. This allows us to transfer the problem (5) into the finite-dimensional form.

We will construct the minimization of the functional (5) on the basis of conjugate gradient method:

$$u^{n+1} = u^n - \gamma_n \cdot s^n \qquad (8)$$

$$s^{n+1} = G'^{n+1} + \beta^n \cdot G'^n \qquad (9)$$

$$\beta^0 = 0 \qquad (10)$$

$$\beta^n = \frac{\left(G'^n, G'^{n+1} - G'^n\right)}{\left\|G'^n\right\|^2} \qquad (11)$$

$$\gamma^n = \frac{\sum_{k=1}^{N_t}\int_0^{\tau_f} \vartheta(x_k^t, y_k^t, \tau)\cdot\left(T\left(x_k^t, y_k^t, \tau\right) - T^e\left(x_k^t, y_k^t, \tau\right)\right)d\tau}{\sum_{k=1}^{N_t}\int_0^{\tau_f} \vartheta(x_k^t, y_k^t, \tau)^2 d\tau} \qquad (12)$$

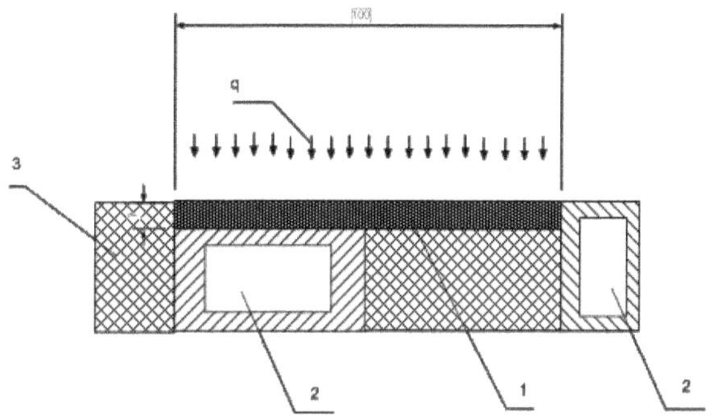

Figure 10. Scheme of the experiment. 1 – sample; 2 – water cooler; 3 – heat insulation. Conditions of the experiment: radiative heat flux heating with the density of 3 MW/m^2; water cooler temperature 300 K; heating duration 300 s.

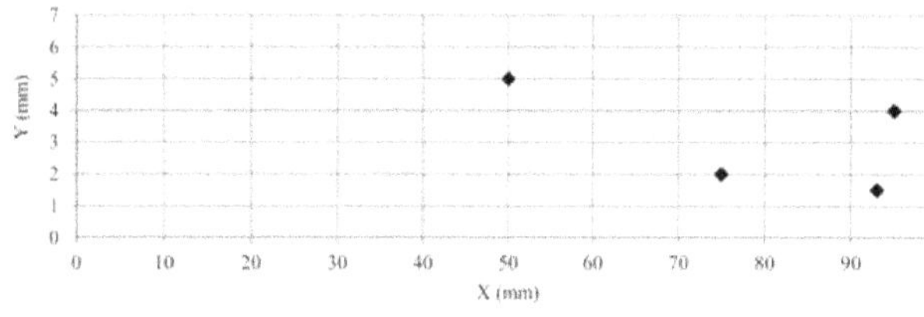

Figure 11. Coordinates of temperature sensors obtained as a result of solving experiment planning problem.

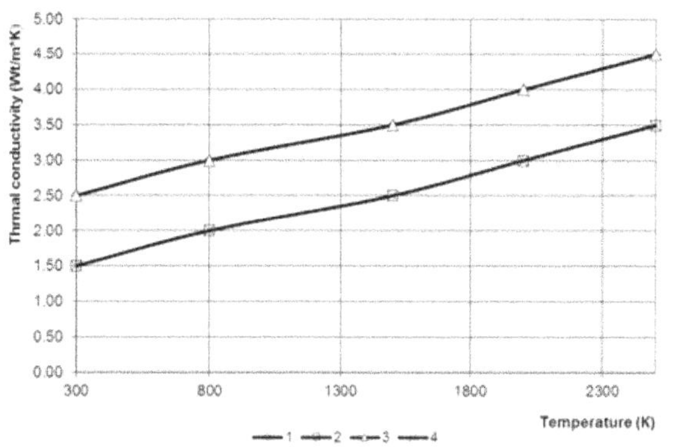

Figure 12. Solution results of the heat conductivity inverse problem for defining the temperature dependencies of thermal conductivity coefficient for the anisotropic material. 1 – dependence $\lambda_x(T)$ of the model material; 2 – dependence $\lambda_y(T)$ of the model material; 3,4 – dependences $\lambda_x(T)$ and $\lambda_y(T)$, obtained from the inverse problem solution.

Functions of sensitivity ϑ are determined from the solution of the following boundary problem:

$$\frac{\partial\left(C(T)\,\vartheta(x,y,\tau)\right)}{\partial\tau}=\frac{\partial}{\partial x}\left(\frac{\partial\left(\lambda_x(T)\,\vartheta(x,y,\tau)\right)}{\partial x}\right)+$$
$$+\frac{\partial}{\partial y}\left(\frac{\partial\left(\lambda_y(T)\,\vartheta(x,y,\tau)\right)}{\partial y}\right)+\frac{\partial}{\partial x}\left(\delta\lambda_x\frac{\partial T}{\partial x}\right)+ \tag{13}$$
$$+\frac{\partial}{\partial y}\left(\delta\lambda_y\frac{\partial T}{\partial y}\right)$$

$$(x,y)\in\Omega \quad \tau\in\left]0,\tau_f\right] \tag{14}$$

$$\tau=0 \quad \vartheta(x,y,0)=0 \tag{15}$$

$$\partial\Omega_1\cap\partial\Omega \quad \vartheta(\partial\Omega_1,\tau)=0 \tag{16}$$

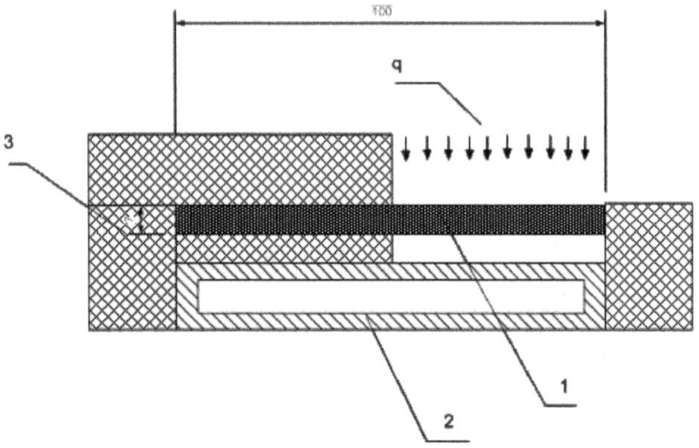

Figure 13. Scheme of the experiment. 1 – sample; 2 – water cooler; 3 – heat insulation. Conditions of the experiment: radiative heat flux heating with the density of 3 MW/m^2; water cooler temperature 300 K; heating duration 300 s.

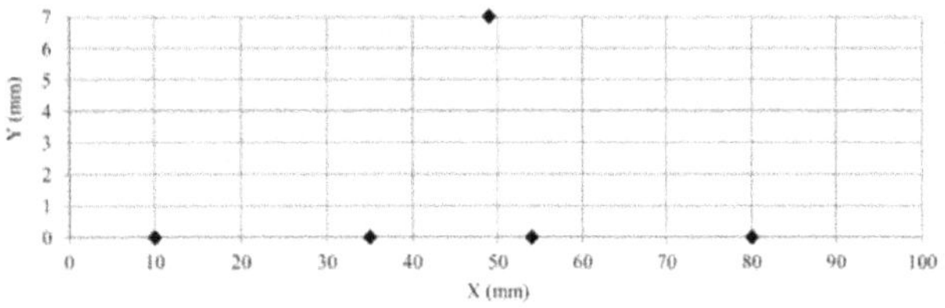

Figure 14. Coordinates of temperature sensors obtained as a result of solving experiment planning problem.

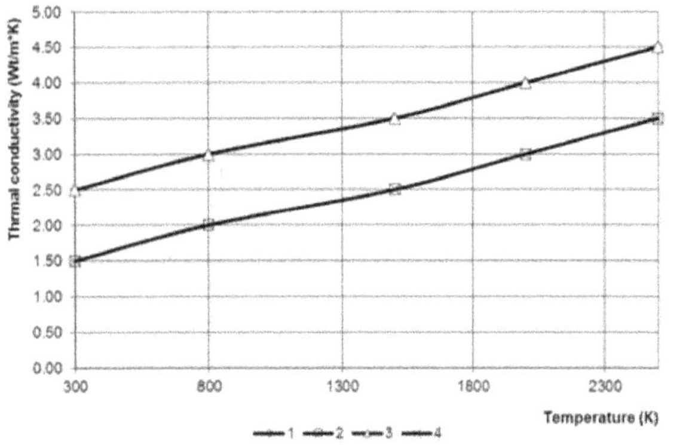

Figure 15. Solution results of the heat conductivity inverse problem for defining the temperature dependencies of thermal conductivity coefficient for the anisotropic material. 1 – dependence $\lambda_x(T)$ of the model material; 2 – dependence $\lambda_y(T)$ of the model material; 3,4 – dependences $\lambda_x(T)$ and $\lambda_y(T)$, obtained from the inverse problem solution.

$$\partial\Omega_2\cap\partial\Omega \quad -\frac{\partial\left(\lambda(T)\cdot\vartheta(\partial\Omega_2,\tau)\right)}{\partial n}=$$
$$=\left[A'_T(T)q_w(\tau)-\varepsilon'_T(T)\sigma_0\left(T^4(\partial\Omega_2,\tau)-T_f^4\right)-\right.$$
$$-\,4\varepsilon(T)\sigma_0T^3(\partial\Omega_2,\tau)-\alpha_f(T) \tag{17}$$
$$\left.-\,\alpha'_{fT}(T)\left(T(\partial\Omega_2,\tau)-T_f\right)\right]\vartheta(\partial\Omega_2,\tau)+$$
$$+\,l_x\delta\lambda_x\frac{dT}{dx}+l_y\delta\lambda_y\frac{dT}{dy}$$

Here l_x and l_y are the cosines of the angle between the inner normal to the area boundary and the coordinates, $\delta u=\{\delta\lambda_x,\delta\lambda_y\}$, $\delta u=s^n$.

The following conjugate problem is defined in order to determine the gradient of discrepancy functional according to the parameters of the required temperature dependences:

$$-C(T)\frac{\partial\psi(x,y,\tau)}{\partial\tau}=\lambda_x(T)\frac{\partial}{\partial x}\left(\frac{\partial\psi(x,y,\tau)}{\partial x}\right)+$$
$$+\lambda_y(T)\frac{\partial}{\partial y}\left(\frac{\partial\psi(x,y,\tau)}{\partial y}\right)+\sum_{n=1}^{N_t}\left[T(x'_n,y'_n,\tau)-\right. \tag{18}$$
$$\left.-T^e(x'_n,y'_n,\tau)\right]\cdot\delta(x-x'_n)\cdot\delta(y-y'_n)$$

$$(x,y)\in\Omega \quad \tau\in\left[0,\tau_f\right[\tag{19}$$

$$\tau=\tau_f \quad \psi(x,y,\tau_m)=0 \tag{20}$$

$$\partial\Omega_1\cap\partial\Omega \quad \psi(\partial\Omega_1,\tau)=0 \tag{21}$$

$$\partial\Omega_2\cap\partial\Omega \quad -\lambda(T)\frac{\partial\psi(\partial\Omega_2,\tau)}{\partial n}=$$
$$=\left[A'_T(T)q_w(\tau)-\varepsilon'_T(T)\sigma_0\left(T^4(\partial\Omega_2,\tau)-T_f^4\right)-\right.$$
$$-\,4\varepsilon(T)\sigma_0T^3(\partial\Omega_2,\tau)-$$
$$\left.-\,\alpha_f(T)-\alpha'_{fT}(T)\left(T(\partial\Omega_2,\tau)-T_f\right)\right]\cdot\psi(\partial\Omega_2,\tau)$$

where: $A'_T=\dfrac{dA}{dT}$, $\varepsilon'_T=\dfrac{d\varepsilon}{dT}$, $\alpha'_{fT}=\dfrac{d\alpha_f}{dT}$.

While the gradients of discrepancy functional can be calculated the following way:

$$G'_{\lambda_{xi}}=\int_0^{\tau_f}\int_\Omega\frac{dT}{dx}\frac{d\psi}{dx}\chi_{1i}(T)\,dxdyd\tau;\; i=\overline{1,K_{\lambda x}} \tag{22}$$

$$G'_{\lambda_{yj}}=\int_0^{\tau_f}\int_\Omega\frac{dT}{dy}\frac{d\psi}{dy}\chi_{2j}(T)\,dxdyd\tau;\; j=\overline{1,K_{\lambda y}} \tag{23}$$

$$G'=\{G'_{\lambda_i},G'_{\lambda_j}\};\, i=\overline{1,K_{\lambda x}};\, j=\overline{1,K_{\lambda y}}$$

4. Temperature measurement planning in researching thermophysical properties of anisotropic material

The approach to temperature measurement planning for thermophysical research was suggested by Alifanov and Artukhine (Artukhine, 1985, Alifanov et al 1988), and was developed in the current paper for two-dimensional heat exchange processes.

By the optimal measurement plan

$$\Lambda=\{N_t,(x_1,y_1),(x_2,y_2),(x_3,y_3)\ldots(x_{N_t},y_{N_t})\}$$

we mean the total of N_t number and the installation coordinates of temperature sensors which ensure the most accurate solution of the inverse problem. As shown in (Artukhine, 1985) for the optimal measurement plan the determinant of normalized Fisher matrix F reaches its maximum: $\Lambda : \max(\det(F(\Lambda))$.

Fisher matrix elements are defined as follows:

$$F(\Lambda) = \frac{1}{N_t}\Phi_{jk};\ j,k = \overline{1,K}$$

$$K = K_{\lambda_x} + K_{\lambda_y} \tag{24}$$

$$\Phi_{jk} = \sum_{n=1}^{N_t}\int_0^{\tau_f}\theta_j(x_n,y_n,\tau)\cdot\theta_k(x_n,y_n,\tau)\,d\tau$$

The following problem is solved in order to determine sensitivity functions $\theta_k(x,y,\tau)$:

$$\frac{\partial\left(C(T)\,\theta_k(x,y,\tau)\right)}{\partial\tau} = \frac{\partial}{\partial x}\left(\frac{\partial\left(\lambda_x(T)\,\theta_k(x,y,\tau)\right)}{\partial x}\right) + \frac{\partial}{\partial y}\left(\frac{\partial\left(\lambda_y(T)\,\theta_k(x,y,\tau)\right)}{\partial y}\right) + Q_k; \tag{25}$$

$$(x,y)\in\Omega,\quad \tau\in\left]0,\tau_f\right],\quad k = \overline{1,K_{\lambda x}+K_{\lambda y}} \tag{26}$$

$$\tau = 0 \quad \theta_k(x,y,0) = 0 \tag{27}$$

$$\partial\Omega\in\partial\Omega_1 \quad \theta_k(\partial\Omega_1,\tau) = 0 \tag{28}$$

$$\partial\Omega\in\partial\Omega_2 \quad -\frac{\partial\left(\lambda(T)\cdot\theta_k(\partial\Omega_2,\tau)\right)}{\partial n} = \left[A'_{w_T}(T)q_w(\tau) - \varepsilon'_{w_T}(T)\sigma_0\left(T^4(\partial\Omega_2,\tau)-T_f^4\right) - 4\varepsilon_w(T)\sigma_0T^3(\partial\Omega_2,\tau) - \alpha_f(T) - \alpha'_{f_T}(T)\left(T(\partial\Omega_2,\tau)-T_f\right)\right]\cdot\theta_k + q_{w_k} \tag{29}$$

$$Q_k = \begin{cases} \dfrac{\partial}{\partial x}\left(\chi_{1k}\dfrac{\partial T}{\partial x}\right), & k = \overline{1,K_{\lambda x}} \\ \dfrac{\partial}{\partial y}\left(\chi_{2k}\dfrac{\partial T}{\partial y}\right), & k = \overline{1+K_{\lambda x}, K_{\lambda x}+K_{\lambda y}} \end{cases} \tag{30}$$

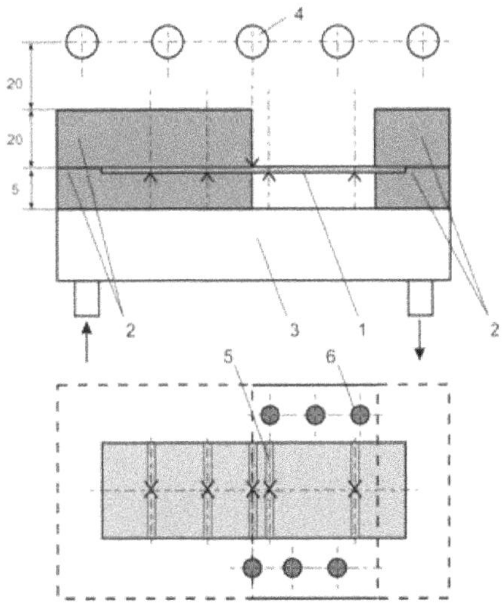

Figure 16. Scheme of carbon-carbon composite material sample tests at UTRO-6 unit. 1 – sample; 2 – protective heat insulation; 3 – water cooler;4 –tungsten halogen lamp; 5 – thermocouple; 6 – heat flux sensor

$$q_{w_k} = \begin{cases} l_x\chi_{1k}\dfrac{dT}{dx}, & k = \overline{1,K_{\lambda x}} \\ l_y\chi_{2k}\dfrac{dT}{dy}, & k = \overline{1+K_{\lambda x}, K_{\lambda x}+K_{\lambda y}} \end{cases} \tag{31}$$

Nelder-Mead non-gradient method is used in order to calculate the maximum of Fisher matrix determinant.

All the boundary problems under consideration are solved numerically with the use of finite-element method.

5. Numerical simulation results

In order to perform a grounded choice of the experimental scheme for studying thermophysical properties of the anisotropic material a series of computational experiments were conducted. At the first stage of each experiment the problem of temperature measurement optimal planning was solved. The number of temperature sensors was taken as equaling four because further increase of their number lead to the placement of two or more sensors in points with close (with the difference not more than 0.5 mm) coordinates which means that the optimal sensor number had been achieved. After that, at the second stage, the problem of direct heat conductivity was solved. Temperature values in the probable points of sensor placement were calculated. These so to say "experimental" temperature values were in their turn used at the third stage in order to solve the problem of inverse heat conductivity for identifying temperature dependences of heat conductivity coefficient and the anisotropic material. Inverse problem solution results vividly show the maximum potential of the abovementioned experimental scheme in the research of thermophysical properties.

The first scheme (Fig. 1) presupposes the use of direct resistance heating. Among the advantages that can be named are its simplicity and low requirement of electric power (not more than 1000 W). However there still exist some problems with isolating temperature sensors from the sample as well as with the isolation of the sample from the test stand construction. In addition as the solution results of temperature planning problems show (Fig. 2) preferred spots for sensor installation are located at the boundaries of the sample which simplifies the installation of sensors. Solution results of heat conductivity inverse problem (Fig. 3) show that in this case one can count on achieving high accuracy in defining temperature dependence of thermal conductivity coefficient in the longitudi-

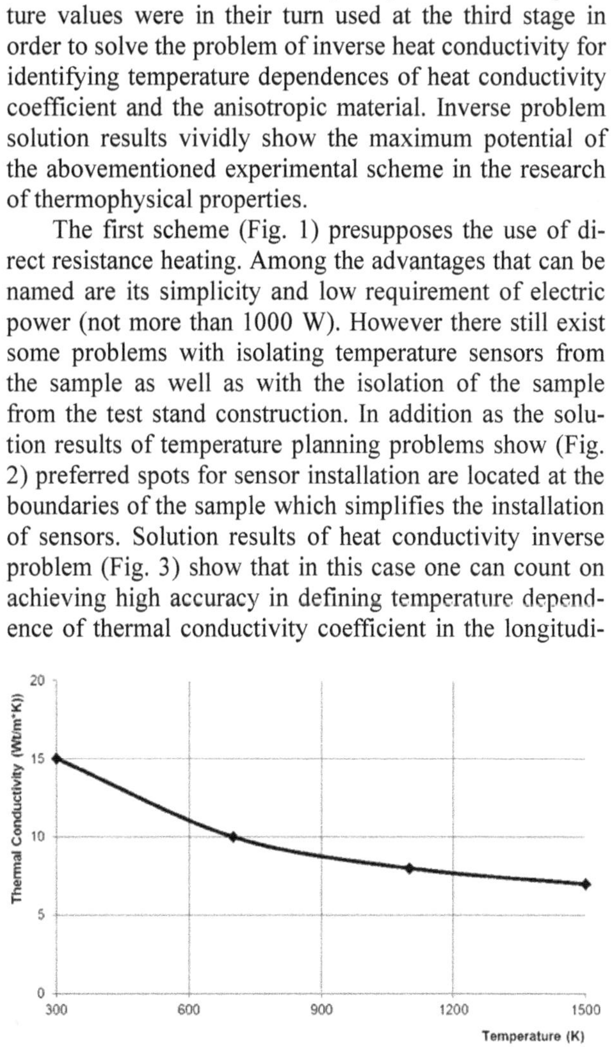

Figure 17. Temperature dependence of thermal conductivity coefficient for carbon-carbon material.

nal direction λ_x in the whole temperature range, and on the recovery of temperature dependence for heat conductivity coefficient in transverse direction λ_y up to the temperature of 1500 K. It is accounted by the fact that temperature sensors which are most informative for defining λ_y are located in a relatively "cold" part of the sample.

The next experimental scheme presupposes the use of concentrated heating sources (heating zone 20 mm), such as plasmatron or units like "Uran" (Fig. 4). Water coolers are used for increasing temperature gradients. It can be seen by the inverse problem solution results that this scheme ensures good results of defining temperature dependences for heat conductivity coefficient in both directions and can be used for conducting experimental research.

The next schemes of experimental research (Fig. 7 and 10) focus on the use of halogen or arc lamp units as heating source which ensures uniform heating of the sample surface. The scheme in Fig. 10 differs by the presence of a side water cooler used to intensify the process of heat conductivity in the longitudinal direction. It should be noticed that the optimal placement of temperature sensors for these schemes is located within the volume of the sample (Fig. 8 and 11). As seen from the inverse problem solution results (Fig. 9) in this case recovery of thermal conductivity coefficient in the longitudinal direction λ_x is absolutely unsatisfactory, however the dependence of thermal conductivity coefficient in the transverse direction λ_y is defined better than in the case with direct resistance heating. Such results occur due to a small temperature difference in the longitudinal direction. The use of an additional side water cooler makes it possible to improve the situation (Fig. 12) and to obtain good results for thermal conductivity coefficient on two directions.

Among the disadvantages of the scheme (Fig. 10) is the need of two water coolers. In order to simplify the experimental method there was considered a case with creating non-uniform heated areas due to "shading" some areas of the sample surface (Fig. 13). This scheme is characterized by the optimal placement of temperature sensors on the sample surface which significantly simplifies the process of their installation (Fig. 14). The results (Fig. 15) show stable identification of temperature dependences for the thermal conductivity coefficient in both directions.

5. Experimental study

This very scheme was used for conducting experimental research at the radiative heating unit Utro-6 (Fig. 16). Tungsten halogen lamps were used as the heating source. Front and back surfaces of the sample were protected by highly porous quartz fiber isolation.

Fig. 17 presents results of obtaining thermal conductivity coefficient of a carbon-carbon material in the longitudinal direction.

References

Reznik, S. et al. (2002). *Materials and Coats in Extreme Conditions. Glance into the Future*. In 3 vol. Moscow: Bauman MSTU Publishing House. (in Russian)

Alifanov, O.M., Vabischevitch, P.N., Mikhailov V.V. et al. (2001). *Identification and Design Foundations of Thermal Processes and Systems*. Moscow: Logos. (in Russian)

Osipova, V.A. (1979). *Experimental Research of Heat Transfer*. Moscow: Energiya. (in Russian)

Filippov, L.P. (1984). *Measurement of Thermophysical Properties by Cyclic Heating Method*. Moscow: Atomenergoizdat. (in Russian)

Platunov, E.S., Buravoi, S.E., Kurepin, V.V. et al. (1986). *Thermophysical Measurements and Devices*. Leningrad: Mashinostroyeniye. (in Russian)

Alifanov,O.M., ArtukhineE.A. and Rumyantsev S.V. (1988). *Extreme Methods of Solving Ill-Conditioned Problems and their Application to Inverse Problems*. Moscow: Nauka. (in Russian)

Artukhine, E.A. (1985) Experimental design of measurement for the solution of coefficient-type inverse conduction problem, *Journal of Engineering Physics and Thermophysics*, Vol. 48, No 3. pp. 372–376.

The Improvement Potential of Friction Surfacing Processes for the Production of Metal Composites

Michael J.R. Stegmüller[1, 2], Richard J. Grant[2], Paul Schindele[1], Zoubir Zouaoui[2]

[1] Laboratory for Welding Techniques, Institute of Mechanical Engineering, Kempten University of Applied Sciences, Bahnhofstraße 61, Kempten, 87435, Germany
[2] Department of Engineering and Applied Physics, Glyndŵr University, Plas Coch, Mold Road, Wrexham, LL11 2AW, United Kingdom

Abstract: The classical solid-state process of friction surfacing offers a large potential for optimization. This paper shows preliminary investigations into two ways in which single layered friction coatings can be improved. Firstly, the classical approach to friction surfacing was supplemented by additional heat sources. Additional heating of the coating material offers the possibility to enhance the number of material combinations. It was possible to produce mild steel, stainless steel and cold work steel coatings on aluminium substrates. Mild steel substrates could be coated with mild steel, stainless steel, cold work steel and chrome-cobalt alloys. Secondly, an alternative method of using a rotating disc as opposed to a rotating rod is shown to coat mild steel ring shaped substrates with aluminium.

Key Words: Improvement potentials, Additional heating, Friction surfacing, Stainless steel, Aluminium, Coating.

1. Introduction

Derived from friction welding, a new process was designed which is widely mentioned in the literature as "friction surfacing". With friction surfacing the face of a rotating cylindrical rod is axially pressed against a flat substrate. The rotating consumable rod is then, under the creation of a flash at the face of the rod, moved along the substrate with a defined feed rate producing a coated layer (Fig. 1). This classical and widely investigated process uses frictionally generated heat which is a function of axial pressure and rotation, for softening both coating and substrate material. However, this classical approach caused problems.

The amount of friction generated heat is a function of the friction coefficient between coating and substrate materials; the friction coefficient must be of a certain value to produce enough frictional heat for the softening of both consumable rod and substrate. Also, the coefficient of friction is a function of the temperature of both coating and substrate material and changes its value during the process of friction coating. Additionally, it is unlikely that hard materials such as steel will coat onto soft substrates such as aluminium because the soft material will deform preferentially (Stern, 1996). In the classical approach a retreating and advancing side (Fig. 1, addition and subtraction of velocities of rotational speed and feed rate) causes an asymmetric material flow and bonding failures at the retreating side (Batchelor, 1996; Yamashita, 2001).

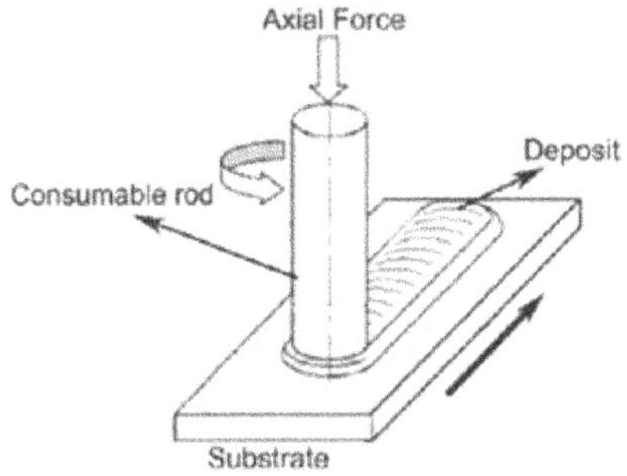

Figure 1. Schematic drawing of friction surfacing process (Khalid, 2012).

As a result, the preliminary stages of this research project offer a scientific investigation concerning the improvement potential of what may be considered as the *classical process* and its associated problems. The use of additional heat sources and their influence on the weldability of material combinations, that can not deliver the desired friction coefficient for producing enough frictional heat to plasticise both coating and substrate material, is subject to closer investigations. Besides, a new approach to the production of friction coatings is presented which concerns the problem of the retreating and advancing side of the consumable rod. This approach (see Fig. 2) may be introduced by the use of the circumferential surface of a rotating disc or rod which provides the coating material and which is pressed onto a flat or round substrate. The disc is moved along the substrate resulting in a friction coated layer. This paper presents the preparatory work leading to initial trials utilizing various approaches and some metallurgical investigations.

2. Experimental setup

The disc experiment was set up on a conventional lathe (see Fig. 3) with a special tool providing the radial clamping of the substrate ring (1) in the chuck while the coating rod (2) is fixed in the collet chuck (3) of an electric drive which can be traversed via a pneumatically traversed carriage (4) with four bearings pressing the coating road onto

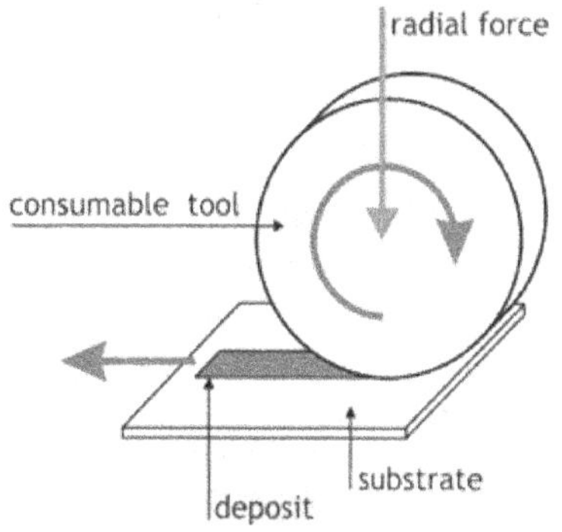

Figure 2. Test arrangement for using the circumferential surface of disc shaped coating material.

Figure 3. Experimental setup for circumferential coating of ring-shaped substrates: a) before trial, b) during friction coating.

the ring (substrate). The softening of the ring and rod is realized by an acetylene torch with less oxygen than acetylene (reduced gas mix). The coating rod and substrate ring can be rotated in the opposite or same direction.

The other technique was approached through the modification of a milling machine (see Fig. 4) which allows fast and easy coating of flat specimens applying a rotating rod. In this case, the cross-section of the rod contacts the flat substrate, and not its circumferential surface. For the first time an inductor coil (4) is used for heating up the coating material (2) before and during the coating of the substrate (1). A pneumatic cylinder is used, pressing the carriage with the electric drive (3) and the fixed coating rod against the substrate.

The first parametric study was conducted by coating flat aluminium substrates (AlMgSi0.5), plate dimensions 20x6.1x135 mm, with stainless steel (X5CrNi18-10), consumable rod dimensions 10 x 96 mm, using this modified milling machine. Such a material combination was thought suitable for this initial study because the union of these and other similar materials offer a number of engineering benefits such as environmental and wear protection. Chandrasekaran et al. (1997a) claimed earlier that this material combination forms an inhomogeneous coated layer with no proper bond; however, initial trials showed otherwise. Consequently, the authors wished to show that stainless steel can be coated onto aluminium forming a consistently good bond with the help of inductive heating.

A separate *start plate*, as suggested by Chandrasekaran et al. (1997a), was manufactured out of mild steel (S235JR), dimensions 20x6.1x30 mm: this would provide a point at which the surfacing process would be initiated. Additionally, the substrate and coating material was prepared by an initial degreasing. The experimental friction surfacing machine had been optimized concerning repeatability by setting a mechanical stop to allow for a constant consumable length (ie a shortening by 57 mm in the coating rod). The value for the mechanical stop was determined by preliminary trials with the maximum length dictated by preventing the flash from contacting the collet chuck of the electric drive; this in turn resulted in the maximum coating length.

Measuring equipment was installed to ensure an easy, fast and accurate collection of data; the data (consumption feed rate of coating material which was obtained by taking the derivative of the consumable traverse with respect to time, maximum process temperature and maximum force) was evaluated with the software Diadem (National Instruments). The temperature of the coating rod at the contact zone rod/substrate was measured with a fixed pyrometer. Afterwards, the flash was created and its surface temperature was measured.

The coating material was preheated (induction unit, Eldec, 3 kW output power) to a temperature of about 830 °C, then the coating process was started. During the

Figure 4. Experimental setup: Modified milling machine.

Figure 5. First ring-shaped specimens, aluminium on mild steel substrate.

Figure 6. Specimens, coating of flat mild steel (S235JR) substrates: a) mild steel (S235JR), b) stainless steel (X5CrNi18-8), c) cold work steel (100Cr6) and d) chrome-cobalt alloy (Stellite 21).

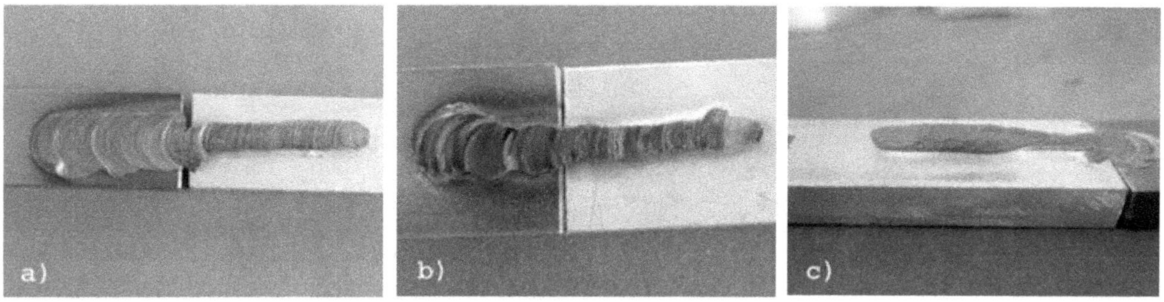

Figure 7. Specimens with start plates: a) mild steel (S235JR), b) stainless steel (X5CrNi18-8) and c) cold work steel (100Cr6) on aluminium substrates (AlMgSi0,5).

coating process the inductive output power was maintained at a constant value, but the inductivity was influenced by the increasing temperature of the coating rod.

After preheating the coating rod to about 830°C it is pressed onto the start plate, with an initial retention/dwell time of 15 seconds before the feed is started. The monitoring of contact pressure, consumption of coating rod and temperature were carried out with the aid of a load cell (manufactured by HBM, Germany), an inductive displacement transducer (HBM) and an infrared pyrometer (Sensortherm). All measured data was recorded by an HBM data acquisition system.

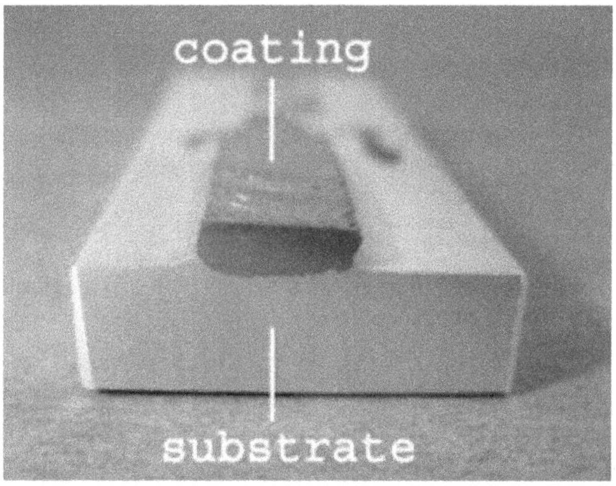

Figure 8. Cross section of cold work steel coating (100Cr6) after cutting: Mechanical interlocking of coating.

3. Results and discussion

3.1 Lathe based approach

The aluminium on a mild steel substrate, using a preheated 20 mm diameter consumable rod, resulted in coatings which were initially judged to be of good quality (Fig. 5). Manual handling and examination seemed to show a good bond strength. Using this lathe technique with other coating materials onto the ring-shaped mild steel substrate met without success; however, other material combinations are subject to future investigations. It should be noted that the coating rod was pre-heated by an acetylene torch.

3.2 Milling machine based approach

Employing the modified milling machine, preliminary trials coating mild steel, stainless steel, cold work steel and a chrome-cobalt alloy on mild steel substrates delivered good coatings with a coating thickness from 1 to 2

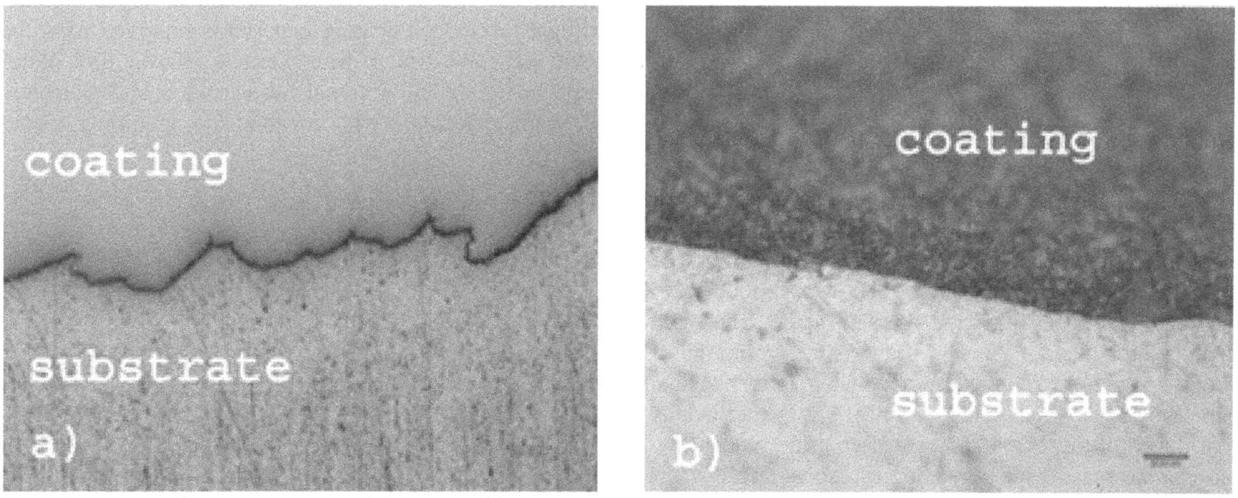

Figure 9. Cross sections of the interface (100Cr6 on aluminium AlMgSi0,5): a) non-etched (50x magnification), b) etched (500x magnification).

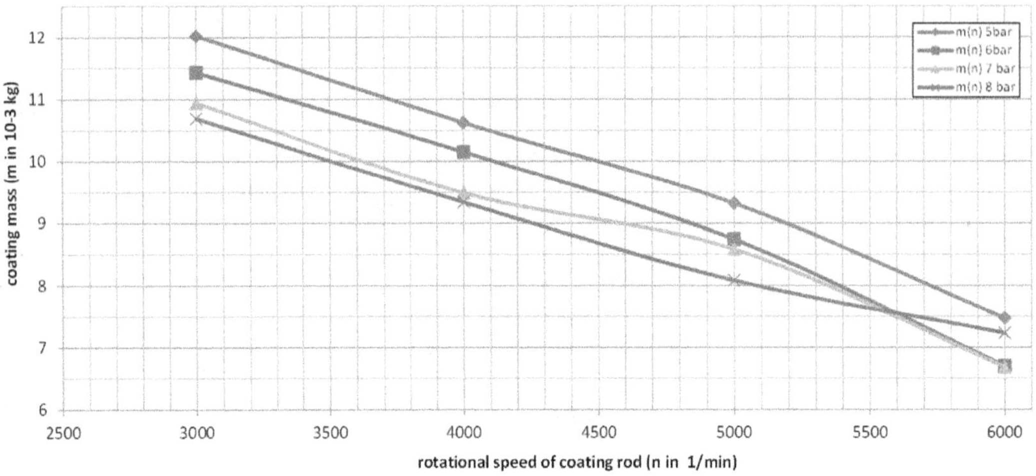

Figure 10. The effect of rotational speed (n) on the coating mass (m), constant pressure (p) from 5 to 8 bar.

mm (Fig. 6). All trials were conducted by applying an inductor coil for heating the coating rod.

The coating of aluminium substrates with hard materials (mild steel, stainless steel and cold work steel, see Fig. 7) can be obtained by using a start plate made out of the coating material. Again inductive heating was used for plasticization of the coating rod.

After transversely cutting the specimens with a high speed cutting disc, mechanical interlocking can be determined with the naked eye (Fig. 8).

Embedding the specimens in plastic, grinding and then polishing followed to allow an investigation with an optical microscope. This visualized the interface between coating and substrate. By way of a comparison, Fig. 9 a shows the mechanical interlocks which were observed by Chandrasekaran et al. (1997b); whereas Fig. 9 b shows, in an etched condition, the microstructure of the hardened coating with its ferrite needles which matches the results of Govardhanet et al. (2012).

The following graphs (Fig. 10 to 13) present the relationship between the coated mass (m) and the input parameters of rotational speed (n) and pressure (p), with a constant feed rate of 150 mm/min. As this work represented an initial investigation the number of specimens was limited. A trial was only repeated when there was concern that the results were suspect: by observation of the test, unexpected values, or clear deviation from trends, and these results were re-evaluated by repetition of the trials. It may be noted that the maximum temperature that could be measured at the surface of the flash produced by the coating rod was approximately 1260°C. Also, the maximum coating rod consumption feed rate that could be determined with about 97 mm/min was at 8 bar pneumatic pressure and 3000 rpm rotational speed.

In Fig. 10 the effect of rotational speed (n) on the coated mass (m) is shown for four different pressure values of the pneumatic cylinder. The diagram shows a decrease in coating mass of about 40% with a doubling of the rotational speed for each pressure value. The maximum coating mass is about 12 grams and the minimum is about 6.7 grams. At 6000 rpm with an interchange of curve positions which suggests a need for further evaluation. Overall, the graph shows a decreasing tendency of the coating mass with increasing rotational speed of the coating rod.

The determination of the flash mass with (4) is calculated by determining the consumed volume in the process which is:

$$V = a \times d^2 \times \pi \times 0.25 \tag{1}$$

where a is the "burn-off length", which means the shortening of the coating rod, and d is the diameter of the coating rod (see Fig. 11).

The volume of the coating (V_c) is:

$$V_c = \frac{m_c}{\rho} \tag{2}$$

where m_c is the mass of the coating and ρ is the density of the coating material.

The volume of the flash (V_F) is determined by the subtraction of the previously volumes calculated:

$$V_F = V - V_C = a \times d^2 \times \pi \times 0.25 - \frac{m_c}{\rho} \tag{3}$$

With constant density of the flash volume the flash mass (m_F) is calculated by

$$m_F = V_F \times \rho = a \times d^2 \times \pi \times 0.25 \times \rho - m_C \tag{4}$$

It can be seen in Fig. 12 which coating mass (m) and flash mass (m_F) (assuming a constant density of the flash material where the flash mass is calculated by (4)) can be achieved for a given rotational speed. The coating mass decreases while the flash mass increases with increasing rotational speed. It is possible that, due to increasing centrifugal forces, the coating material is removed from the rotational contact plane when there is a plastic flow of coating material. The minimal flash mass is approximately 12 gram, calculated with equation (4), assuming a con-

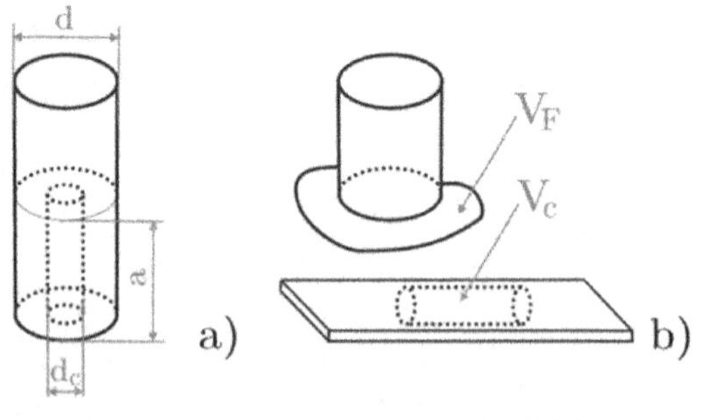

Figure 11. Shortening (a) of the coating rod with diameter of real rotational contact plane (d_c) as determined by Fukakusa (1996): a) before and b) after the friction surfacing process.

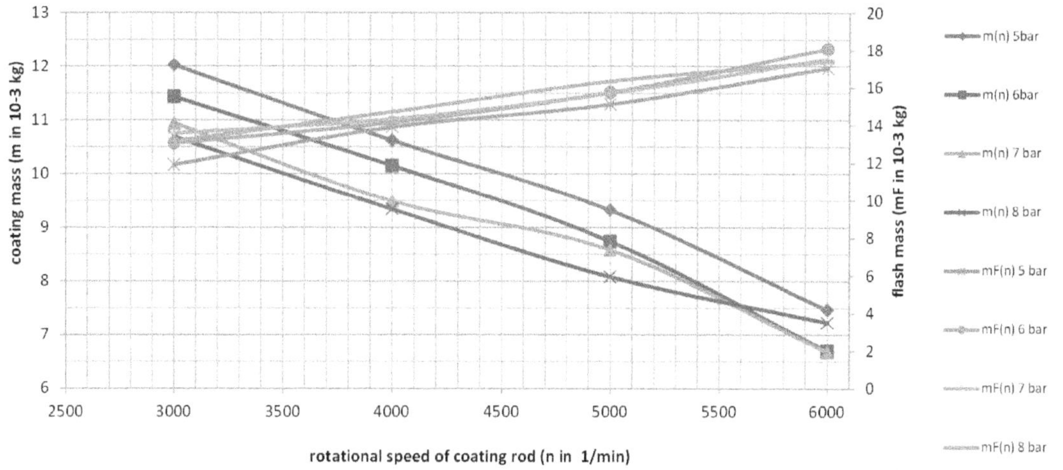

Figure 12. The effect of rotational speed (n) on coating mass (m) and mass of the created flash (mF) from 5 to 8 bar.

stant density of the coating material of 7.9×10^{-3}g/mm^3. This supplements the investigations of Fukakusa (1996).

The graph in Fig. 13 presents the coating mass (*m*) achieved with increasing pressure (*p*) for four rotational speeds. Overall, increasing pressure results in a decreasing coating mass with one exception at the rotational speed of 6000 rpm. As can be seen, the coating mass increases for a decrease in the rotational speed; this is a similar result to that shown in Fig. 10. However, as the slope of the 3000 and 4000 rpm curves is reducing, it could be that for a pressure in excess of 8 bar other rotational speeds might show curves of a concave form – this would require further investigation.

The use of an additional heat source improved the adhesive properties of material combinations that cannot deliver an appropriate friction coefficient necessary for producing enough frictional heat to plasticise both coating and substrate material. It was therefore possible to create new material combinations for novel engineering applications. Material combinations which suggested metallurgically sound bonding (inspection with an optical microscope) were: mild steel, stainless steel, cold work steel, chrome-cobalt alloy on mild steel substrates; mild steel, stainless steel and cold work steel, on aluminium substrate (flat substrates); and aluminium on a mild steel ring shaped substrate. Various coating thicknesses from 0.5 to 3.3 mm were obtained. The principal parametric relationship that was found for the friction surfacing process was that a decrease in coating mass, and an increase in the produced flash mass are apparent for an increase in the rotational speed of the consumable rod. The relation of the process parameters can be compared and coincide with results of other published investigations with various material combinations.

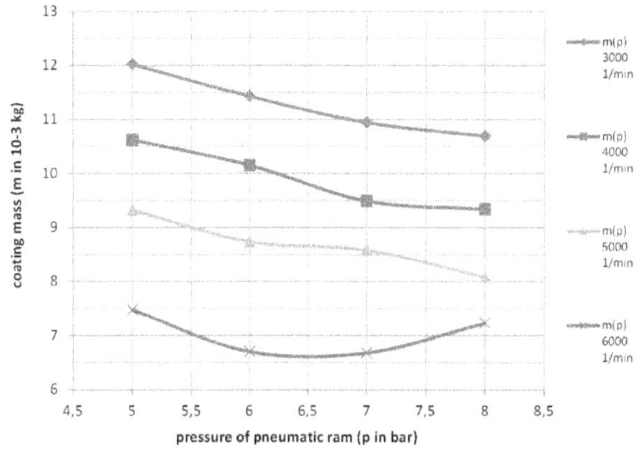

Figure 13. The effect of pressure (p) on coating mass (m), constant rotational speeds (n) from 3000 to 6000 rpm.

4. Outlook

The use of mechanical testing to obtain the relative bond strengths of the specimens has yet to be done. Designing and using an apparatus to gain shear test results is reasonably straightforward. Obtaining absolute values of tensile bond strength is not a simple task when dealing with coating specimens as produced in this work. The completion of preliminary trials and the ongoing improvement of the experimental setups (milling machine and lathe) for using a rod and a disc as coating material has provided useful information to allow the implementation of an appropriate selection of measuring equipment. The opportunity is now to re-run these and similar tests to improve and gain confidence in the data. Further parametric studies will be conducted to collect more process data of other material combinations and hence obtain their substrate-coating bond strengths.

References

Batchelor, A.W., Jana, S., Koh, C.P., and Tan, C.S. (1996). The effect of metal type and multi-layering on friction surfacing, *Journal of Materials Processing Technology*, Vol. 57, No 1-2, pp. 172-181.

Chandrasekaran, M., Batchelor, A. and Jana, S. (1997). Friction surfacing of metal coatings on steel and aluminium substrate, *Journal of Materials Processing Technology*, Vol. 72, No 3, pp. 446-452.

Chandrasekaran, M., Batchelor, A. and Jana S. (1997). Study of the interfacial phenomena during friction surfacing of aluminium with steels, *Journal of Materials Science*, Vol. 32, pp. 6055-6062.

Fukakusa, K. (1996). On the characteristics of the rotational contact plane - A fundamental study of friction surfacing, *Welding International*, Vol. 10, No 7, pp. 524-529.

Govardhan, D., Kumar, A., Murti, K. and Madhusudhan Reddy, G. (2012). Characterization of austenitic stainless steel friction surfaced deposit over low carbon steel, *Materials & Design*, Vol. 36, pp. 206-214.

Khalid Rafi, H., Janaki Ram, G., Phanikumar, G. and Prasad Rao, K. (2011). Microstructural evolution during friction surfacing of tool steel H13, *Materials & Design*, Vol. 32, No 1, pp. 82–87.

Stern, K. H. (1996). *Metallurgical and Ceramic Protective Coatings.* London: Chapman and Hall, pp. 168-193.

Yamashita Y. and Fujita K. (2001). Newly Developed Repairs on Welded Area of LWR Stainless Steel by Friction Surfacing, *Journal of Nuclear Science and Technology*, Vol. 38, No 10, pp. 896-900.

Mass and Cost Effective Wing for Tourist Class Reusable Space Vehicle

Tatiana Ageyeva, Inna Shafikova

Rocket and Spacecraft Composite Structures Department, Faculty of Special Machinery, Bauman Moscow State Technical University, 5 2nd Baumanskaya Street, Moscow, 105005, Russia

Abstract: One-criterion optimization with definite performance criterion is a rather rare thing for the real-world problems. When we deal with the designing of such complex object as reusable space vehicle (RSV) its efficiency hardly can be estimated with only one characteristic. Usually it is necessary to add supplementary criteria (efficiency indexes) which help to obtain fuller information about advantages and disadvantages of the project. As for finding the optimal solution a significant number of different factors should be taken into account during requirements analysis, therefore the design objectives for complex structures are always multicriteria. For instance, such objectives often occur in construction design, integrated circuit design, military science etc.

Key Words: Multicriteria optimization, Hybrid composite, Reusable space vehicle.

1. Problems and solutions of multicriteria optimization

One of the main multicriteria optimization problems is criteria normalization. As different criteria have different physical meaning and units and their scales are incomparable, matching the results for each criterion is impossible. Operation which unifies scales of local criteria is called "criteria normalization". Normalized criterion is a ratio of "natural" particular criteria to normalization constant. The selection of normalization constant must be justified. Several approaches of the selection are described below:

- consumer demands for the product;
- maximum criteria values from the decision region (region D);
- the best achieved parameter value;
- difference between maximum and minimum criteria values in D region;

$$f_i(X) = \frac{F_i^{max} - F_i(X)}{F_i^{max} - F_i^{min}} \quad (1)$$

or

$$f_i(X) = \frac{F_i(X) - F_i^{min}}{F_i^{max} - F_i^{min}} \quad (2)$$

where $f_i(X)$ is normalized particular criterion.

The second issue of multicriteria optimization is the difficulty of trade-off decision searching. One of the solutions is to turn vector optimization problem into parametric optimization, which means one-criterion (scalar) optimization problem.

In other words, all the particular criterion $F_i(X)$ are integrated into complex (generic) criterion $f(X) = \Phi[F_1(X), F_2(X), \ldots, F_m(X)]$ which then is optimized.

Formally generic criterion is a function Φ: $Y_1 \mathrm{x} Y_2 \ldots \mathrm{x} Y_m \rightarrow E$, where Y_j is a set of estimations on j-criteria. As soon as generic criteria Φ is specified for each legitimate result $X \in D$ numeric evaluation of its efficiency $f(X) = \Phi[F_1(X), F_2(X), \ldots, F_m(X)]$ can be found.

Thus generic criterion reduce multicriterion optimization problem to one-criterion with fitness function $f(X)$.

The most common generic criteria are "balanced sum of particular criteria" (or "additive criterion"), "multiplicative criterion" and "ideal point method" (Gorbunov, 2010).

1.1. Additive criterion

The generic criterion is presented as :

$$f(X) = \sum_{i=1}^{m} \lambda_i F_i(X) \quad (3)$$

where λ_i are weight factors, which specify i-criteria preferences in comparison with the other criteria and the value of λ_i defines priority of each criteria. The criteria which is more important is attributed more weight and the sum of all the criteria is assumed to be 1:

$$\sum_{i=1}^{m} \lambda_i = 1 \quad (4)$$

The following are two weight factor definition methods:

- Expert appraisal methods;
- Formal methods.

Expert appraisal methods are very sensitive because choices of decision making persons are based on their personal preferences. Formal method generally is based on the particular criteria value spread and implemented as following sequence of operations:

- Coefficient of relative spread calculation for each particular optimal criterion $F_i(X)$:

$$\delta_i = \frac{F_i^+ - F_i^-}{F_i^+} \quad (5)$$

where $F_i^- = \min F(X)$, $F_i^+ = \max F(X)$, δ_i defines worst possible deviation for the i-criteria.

- Weight factors are defined as follows:

$$\lambda_i = \frac{\delta_i}{\sum \delta_k} \quad (6)$$

1.2. Multiplication criterion

Multiplication criterion is formed by multiplication of particular criteria if they have the same importance. If the particular criteria have different importance weight factors are introduced and multiplicative criterion is defined as follows:

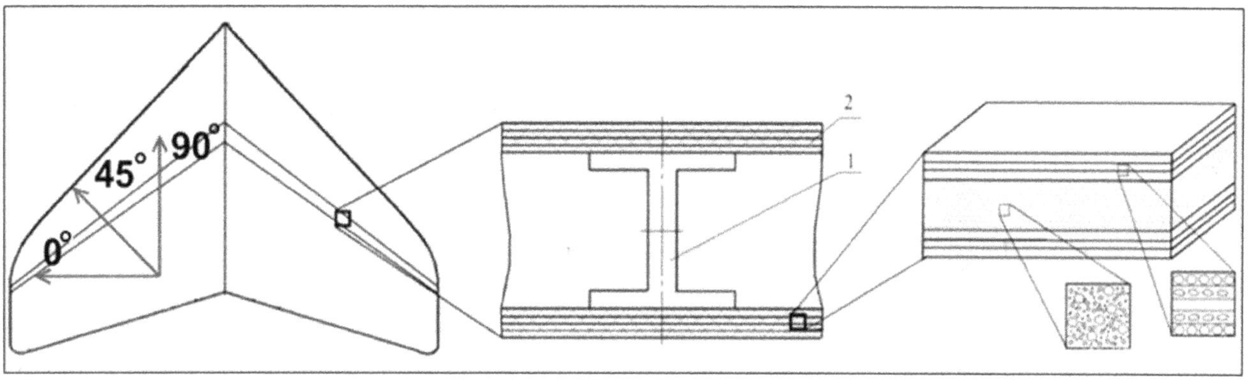

Figure 1. RSV wing structure.

$$F(X)=\prod_{i=1}^{m}F_i^{\lambda_i}(X) \qquad (7)$$

One of the advantages of the multiplicative criterion is that there's no need to normalize particular criteria. The disadvantage is that multiplicative criteria compensate low value of one particular criterion by high value of another and has a tendency to smooth levels of particular criterion at the expense of their unequal primary values.

1.3. Ideal point method

The main idea of the method is that an m-field is observed (where m – is a number of local criteria). In this field a vector illustrating "ideal" solution is selected (e.g. minimal or maximal values of local criterion).

In this field a metric is defined in order to figure out the distance between the decision and "ideal" vectors.

The nearest to "ideal" solution is selected as the best. The drawback of the method is the absence of rules for "ideal" point determination and metric. Fitness function in case of ideal point method is defined as:

$$f(X)=\sum\lambda_i\left(1-f_i(X)\right)^2 \qquad (8)$$

Table 1
Parameters of the materials for the wing

Material physical properties	Unit	
Density of epoxy resin	kg/m^3	1200
Density of GF	kg/m^3	2500
Density of CF	kg/m^3	1800
Resin mass faction	%	30%
GF mass fraction	%	70%
CF mass fraction	%	0%
Density of composite	kg/m^3	2110
Parameters of FRP list		
Thickness of 1 layer of fabric	mm	0.25
Number of fabric layers in 1 list	psc	6
Thickness of 1 list	mm	1.50
Square of 1 list	m^2	16
Volume of 1 list	m^3	0.02
Mass of 1 list	kg	50.64
Density of the list	kg/m^2	3.17
Core parameters		
Core thickness	mm	25.00
Honeycomb density (PAMG-XR1-0015-N-5052)	kg/m^3	97.70
Core mass for 1 skin	kg	39.08
Spar parameters		
Length	m	4.73
Height 1	m	0.40
Flange 1	m	0.25
Thickness 1	m	0.04
Height 2	m	0.20
Flange 2	m	0.10
Thickness 2	m	0.01
Spar volume	m^3	0.07
Spar mass	kg	115.10
Spar material cost	rub	413 242

2. Multicriteria optimization of hybrid composite wing

The wing sweep is 45°, effective airfoil area is 16 m^2. 2 skins (lower and upper) are required for one wing). There are 4 FRP sheets, 2 core layers and one CFRP spar in one wing (Fig. 1). 1 sheet is upper or lower skin of sandwich structure. 1 sheet consists of 6 layers of fabric (either glass or/and carbon). So, FRP for 2 sheets are 8 pieces.

3. Optimization problem

Design factors: number of GFRP and CFRP layers (n_{GFRP}, n_{CFRP}) and fraction of 0°, ±45°, 90° layers (v_0, $v_{\pm45}$, v_{90}).

Optimization criteria (Golushko and Nemirovskiy, 2008) and mass of hybrid composite skin (Pilyugina and Ageyeva, 2012):

$$\overline{M}=\sum_{n=1}^{N}\int_{h_{n-1}}^{h_n}\left(\iint_S\overline{\rho}^{(n)}dS\right) \qquad (9)$$

$$\overline{\rho}^{(n)}=v_m^{(n)}\overline{\rho}_m^{(n)}+\sum_{k=1}^{K^{(n)}}v_f^{(n)}\overline{\rho}_f^{(n)} \qquad (10)$$

$$v_m^{(n)}=1-\sum_{k=1}^{K^{(n)}}v_f^{(n)} \qquad (11)$$

where $\overline{M}$ is mass of hybrid multilayer composite skin; $\overline{\rho}^{(n)}$ is volume density of n-layer; $\overline{\rho}_m^{(n)}$, $\overline{\rho}_f^{(n)}$ are volume

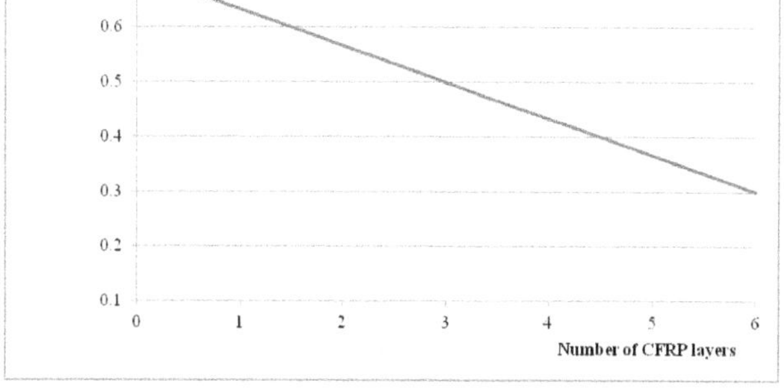

Figure 2. Additive fitness function.

Table 2
Example of materials cost calculation

Cost parameters							
Materials for 1 list (glass)		Waste, %	Unit	Materials for 1 list	Material cost, rub/kg (GBP/m^2)	Cost, GBP	Cost of 1 list, GBP
Resin (Huntsman)	epoxy	1.00%	kg	8.6	6.2	54	135
Glass fabric Ortex 720 1000 (Owens Corning)		1.00%	m^2	96.0	0.8	81	
Materials for 1 list (carbon)							
Resin (Huntsman)	epoxy	1.00%	kg	8.6	6.2	54	2870
Carbon fabric (280 g/m^2), (HC Composite, Russia)		1.00%	m^2	96.0	29	2816	

Table 3
Mass and cost of the wing in accordance with the different variants of hybrid composite package

Hybrid composite				
Number of GFRP layers, pcs	Number of CFRP layers, pcs	Mass of the list, kg	Mass of 1 wing, kg	Material cost for one wing, GBP
6	0	50.64	318	9 107
5	1	48.68	310	10 913
4	2	46.72	302	12 719
3	3	44.76	294	14 525
2	4	42.80	286	16 330
1	5	40.84	278	18 136
0	6	38.88	271	19 942

densities of matrix and fibers of K-family in n-layer; $K^{(n)}$ is the number of fibers families in n-layer; $v_m^{(n)}$, $v_f^{(n)}$ are matrix and fiber fraction in n-layer.

Material cost is defined as follows:

$$\overline{P} = \sum_{n=1}^{N} \int_{h_{n-1}}^{h_n} \left(\iint_S \overline{p}^{(n)} dS \right) \quad (12)$$

$$\overline{p}^{(n)} = v_m^{(n)} \overline{\rho}_m^{(n)} \overline{P}_m^{(n)} + \sum_{k=1}^{K^{(n)}} v_f^{(n)} \overline{\rho}_f^{(n)} \overline{P}_f^{(n)} \quad (13)$$

where P is materials cost for the wing skin; $\overline{P}^{(n)}$ is relative cost of n-layer; $\overline{P}_m^{(n)}$, $\overline{P}_f^{(n)}$ are relative costs of matrix and K-family fiber costs in n-layer.

Wing deflection limitation is:

$$\overline{J} = W(x^*) \quad (14)$$

$$\overline{J} = W_0(x^*) \quad (15)$$

$$J = \frac{\overline{J}}{\overline{J}_0} \quad (16)$$

$$(x^*) \in S \quad (17)$$

where (x^*) is the coordinate of maximal deflection; $\overline{J}$ $\overline{J}_0$ – deflection vectors of the wing and "ideal" wing.

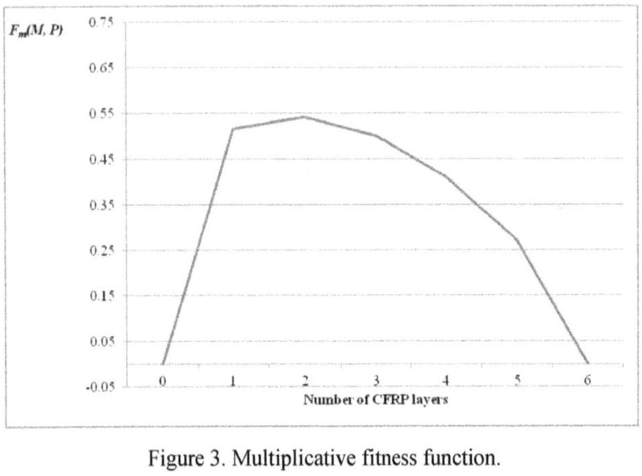

Figure 3. Multiplicative fitness function.

Scalar function which must be minimized is a set of vectors $\overline{M}, \overline{C}, \overline{J}$.

Further the selection of appropriate form of scalar function will be done.

4. Selection of scalar function

To simplify the task it is possible to optimize the structure only for 2 parameters: mass and cost. Then the optimization problem will be the following (Reznik, Prosuntsov and Ageyeva, 2013; Ageyeva and Reznik, 2011):

$$F(M,P) \rightarrow min \quad (18)$$

In accordance with the described algorithm of reducing the vector optimization problem to the scalar problem there are 3 variants of building up fitness function: "additive", "multiplicative" and "ideal point". Input data for the function is shown in table 3.

Normalized mass and cost criterion are defined according to formula (1).

M_{norm}	P_{norm}
0	1.00
0.16	0.83
0.33	0.66
0.50	0.50
0.66	0.33
0.83	0.16
1.00	0

Definition of relative spread coefficient and weight factors (formulas 3, 4):

δ_1	0.23
δ_2	0.54
λ_1	0.30
λ_2	0.70

where δ_1 and δ_2 are coefficients of relative spread for correspondingly mass and cost, λ_1 and λ_2 are weight factors for mass and cost.

5. Additive criteria $F_a(M, P)$

$$F_a(M,P) = 0.3M_{norm} + 0.7P_{norm} \quad (19)$$

Analysis of additive fitness function shows that its optimization (with selected weight factors) will bring it to the obvious result – fully carbon wing.

6. Multiplication criteria $F_M(M, P)$

$$F_M(M,P) = M_{norm}^{0.3} P_{norm}^{0.7} \quad (20)$$

So, according to figure 3, multiplicative fitness function will bring to obvious results as well as additive – to fully GFRP and fully CFRP wing.

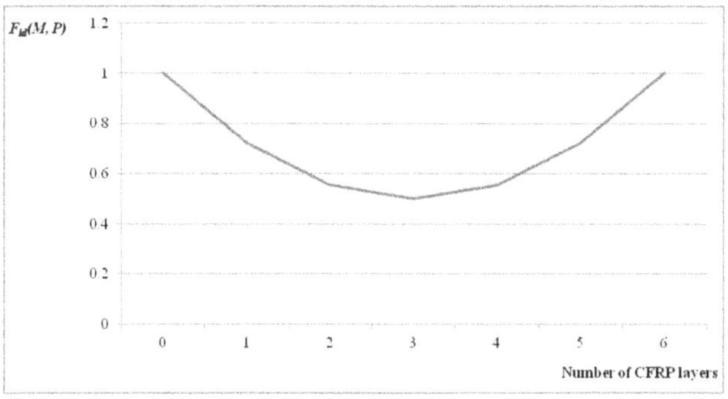

Figure 3. Multiplicative fitness function.

7. Ideal point criteria $F_{id}(M, P)$

$$F_{id}(M,P) = (1 - M_{norm})^2 + (1 - P_{norm})^2 \quad (20)$$

8. Conclusion

The most suitable fitness function for the current task (optimization of hybrid composite wing) is "ideal point" fitness function. Depending on the choice of weight factor minimum of the function can remove to the site of reduction/increasing of the number of CFRP layers in the package.

References

Ageyeva, T.G. and Reznik, S.V. (2011). Multidisciplinary optimization of hybrid composite wing structure for reusable space vehicle, in *Proceedings of the 1st International Workshop on Advanced Composite Materials and Technologies for Aerospace Applications*, Wrexham, UK, May 9-11, 2011, Glyndŵr University, pp. 31-42.

Benjamin, J.P. (2009). *Multidisciplinary Optimization of a Carbon Fiber Reinforced Plastic Wing Cover*, PhD Thesis, Cranfield University.

Golushko, S.K. and Nemirovskiy, Y.V. (2008). *Direct and Inverse Problem of Mechanic of Elastic FRP Plates and Rotary Shells*, Moscow: Fizmatlit. (in Russian).

Gorbunov, V.M. (2010). *Decision Making Theory*, Tomsk: National Research Tomsk Polytechnic University.

Pilyugina, A.V. and Ageyeva, T.G. (2012). Technical and economic efficiency of tourist class space vehicles projects, *Bulletin of Bauman MSTU, Series "Machine Building", Special Issue № 3 'Advanced Materials, Structures and technologies for Rocket-Space Engineering industry"*, pp. 107-119 (in Russian).

Reznik, S.V., Prosuntsov, P.V. and Ageyeva, T.G. (2013). Optimal design of the suborbital reusable spacecraft wing made of polymer composite, *Scientific and Technical Journal FGUP NPO n.a. S.A. Lavochkina*, No 17, pp. 38-43 (in Russian).

Development and Simulation of a Thermoforming Validation Process for Textile Thermoplastic Composites

Werner A. Hufenbach, Bernhard Maron, Moritz Weidelt, Albert Langkamp

Institute of Lightweight Engineering and Polymer Technology (ILK), Faculty of Mechanical Science and Engineering, Technische Universität Dresden, Holbeinstraße 3, Dresden, 01307, Germany

Abstract: Textile thermoplastic composites can be processed using highly efficient production technologies. In particular, the thermoforming technique enables short cycle times. Numerical analyses facilitate the development of these production processes and can be used to describe the resulting fibre structure after forming which is important for structural mechanics calculations. The objective of the presented work is the development of a thermoforming process for the validation of Finite Element forming simulations. The experimental setup allows typical process parameters such as temperature, pressure and velocity to be controlled to meet specific material and process requirements. The validation experiments are conducted with multi-layered flat knitted textiles made of glass fibre/polypropylene hybrid yarn. Material data required for the simulation model is determined using shear tests, tensile tests, bending tests and friction tests at process temperature. Experimental results are finally compared to simulation results.

Key Words: Space antennas, Composite materials and structures, Metal meshes, Mathematical modelling, Tests.

1. Introduction

Composites made of textile thermoplastics are suitable for industrial series production processes like thermoforming (Hufenbach et al., 2010a). To create an efficient process while fulfilling the high demands on component quality is a complex task. In comparison with well-established pressing processes for metallic materials, only minor experience exists concerning the recent textile thermoforming technology. Exemplary non-destructive studies on a consolidated thermoformed component confirm the occurrence of complex draping mechanisms and textile rearrangements during the thermoforming process (Hufenbach et al., 2011). The fact that the mechanical behaviour of composite materials can be altered on microscopic, mesoscopic and macroscopic level (e.g. fibre, matrix, weave/knit construction, lay-up) results in numerous independent parameters of their forming processes. A substantial development for each new process is the consequence. To avoid time-consuming and costly trial-and-error methods, numerical forming simulations can be applied (Nino et al., 2008). Mechanical properties of the textile which are required as simulation input can be determined by experiments. In the context of this research a thermoforming process and tool are developed which can be utilised for validation of thermoforming simulation models for various textile thermoplastics and process conditions. A hybrid yarn-based flat knitted composite material (Engelmann et al., 2002) is characterised through established and proven tests. The data sets are used for numerical simulations within the commercial Finite Element (FE) package PAM-FORM®. Subsequently, analysis results are validated against results of forming experiments carried out with the newly developed tool.

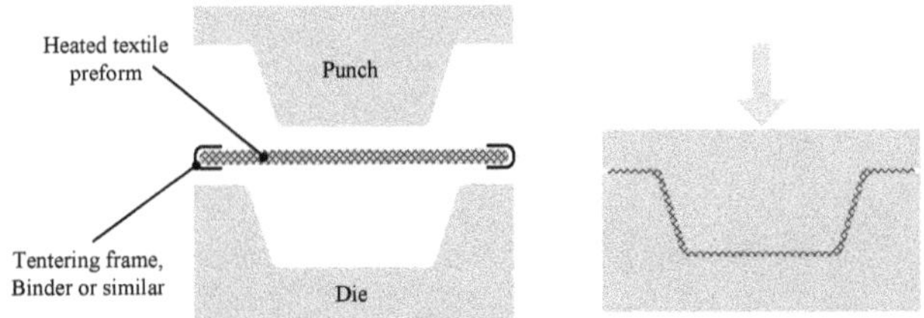

Figure 1. Schematic of a thermoforming process for continuous fibre-reinforced composites

2. Thermoforming process

Industrial thermoforming processes for continuous fibre-reinforced thermoplastics are commonly applied to pre-consolidated composite sheets, also referred to as organic sheets. The hybrid yarn-based textile thermoplastics used within this study are processed without pre-consolidation. Forming, compression and consolidation take place during a single pressing process. During the thermoforming process (Fig. 1) the preform is heated above melting temperature of the matrix and then pressed into a die by a matching punch.

Areas with steep wall sections may suffer from poor consolidation due to the lower pressure. Flexible tools can be used to establish a uniform pressure field (Brooks, 2007). As an example, Fig. 2 shows composite trays made of hybrid yarn-based textile thermoplastics, which were examined within the transfer project T5 of the Collaborative Research Centre SFB 639 at TU Dresden (Hufenbach et al., 2010b, Hufenbach et al., 2009). The tray in Fig. 2a was manufactured using an adapted hot pressing manufacturing scheme with silicone membranes.

Since heating of the preform is by far the most time consuming part of the process, heating and forming are usually allocated to separate stations in order to reduce cycle times. For the configuration of the heating devices, the following aspects are particularly important: energy efficiency, high heating rates and fast transport to the forming station. In this context, infrared radiative heating panels have proven to be suitable for industrial as well as scientific purposes (Pantelakis et al., 2009). For the fixation of textile thermoplastics during transport between the stations and to create a well-defined and reproducible process (e.g. retain fibre architecture, transfer time), different approaches such as grabber systems, tentering frames, spring-loaded blank holders or tool-integrated blank holder systems are used (Lee et al., 2007).

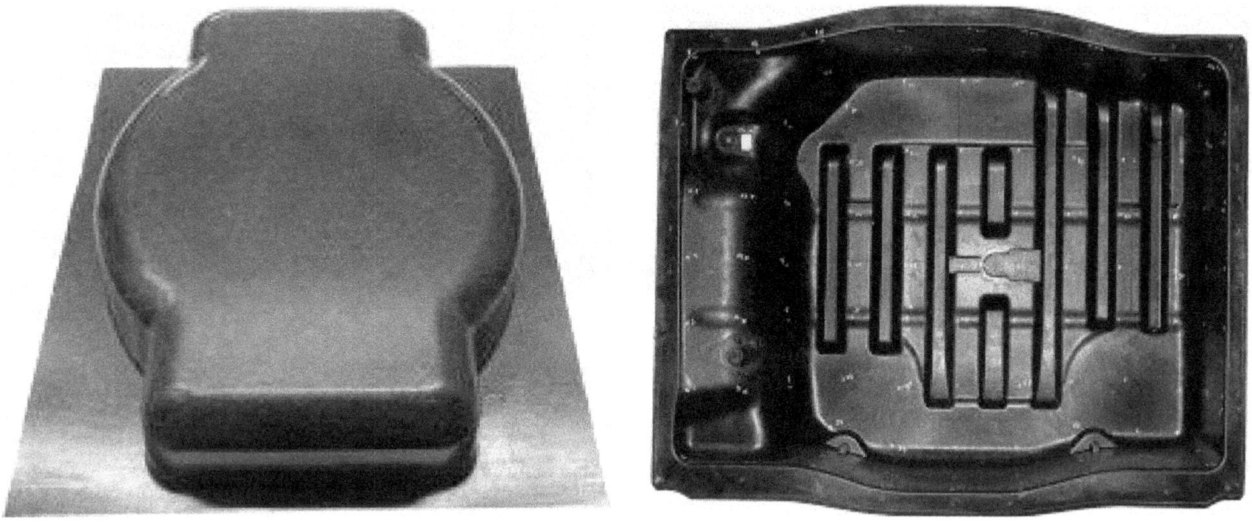

a) b)

Figure 2. Trays made of hybrid yarn-based thermoplastic composite: (a) Tray produced by hot pressing technology, (b) Tray with complex ribs (Hufenbach et al., 2010b).

3. Development of the validation geometry and tool design

Great design freedom is an inherent advantage of the application of composite materials – resulting in products with great geometric complexity. Therefore, validation geometries used in comparable studies often have double-curved surfaces (Khan et al., 2010; Cherouat and Borouchaki, 2009; Willems et al., 2006). A cylindrical cavity was chosen as a starting point for the development of the present validation tool. The final shape was determined with support of kinematic drape modelling (Van Der Weeën, 1991). These simulations are based on a solely geometric approach that helps to predict shear angles within the textile. A net of inextensible fibres is thereby placed on the examined surface and draped around it. Fig. 3 shows an exemplary simulation result and the parameters varied during the geometric optimisation.

The defined demonstrator geometry is incorporated into a thermoforming tool that is composed of the following main components: die, punch adapter with an exchangeable punch and a variable binder system which permits the use of different binder forces and binder section geometries. Due to the abrasive impact of fibre-reinforced composites, die and punch adapter are manufactured from pre-tempered and corrosion-resistant tool steel (X33CrS16). The experiments in this study are conducted with a punch made of conventional high alloy steel (X5CrNi18-10) and a mould cavity (i.e. wall thickness of the composite part) of 1 mm. However, the tool can also be equipped with a flexible punch (e.g. polyurethane, silicone) or a punch which creates a differing mould cavity. The overall dimensions of the closed tool are approx. 400 mm x 300 mm x 240 mm.

An infrared panel heats the composite material to process temperature, while six temperature sensors register the process temperatures of the tool and the composite material. Fig. 4 shows different views of the tool.

4. Material characterisation and simulation

The material used in this study is a multi-layered flat knitted textile (Engelmann et al., 2002). It consists of a knitted textile with additional aligned yarns in the warp and weft direction and was developed at the Institute of Textile Machinery and High Performance Material Technology (ITM) at TU Dresden. It combines excellent draping quality of knitted textiles with the load bearing qualities of non-crimp fabrics. The textile is produced from TWINTEX® material which is a glass fibre/polypropylene hy-

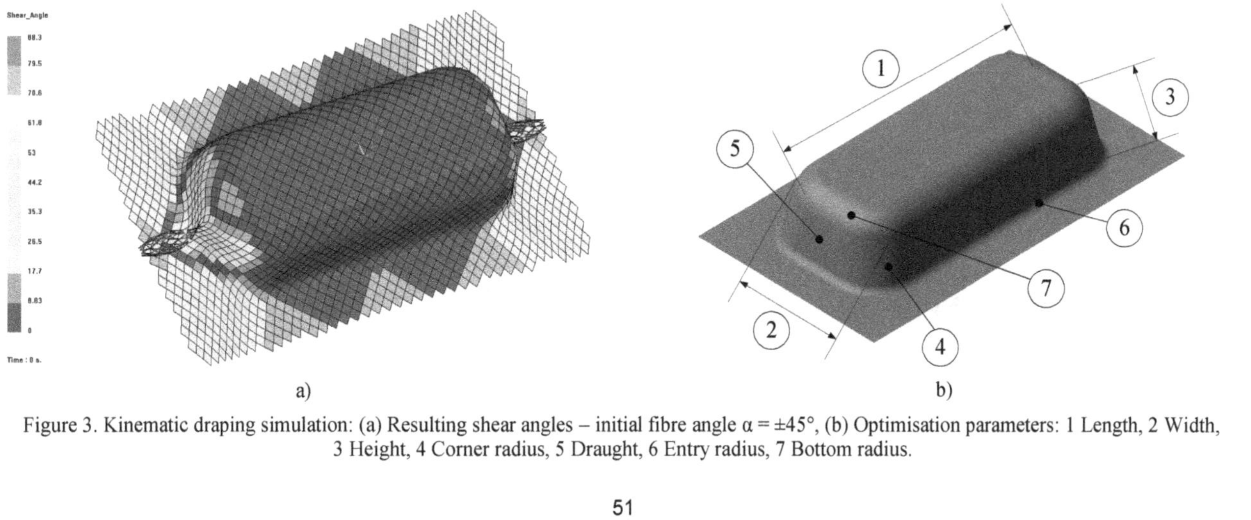

a) b)

Figure 3. Kinematic draping simulation: (a) Resulting shear angles – initial fibre angle $\alpha = \pm 45°$, (b) Optimisation parameters: 1 Length, 2 Width, 3 Height, 4 Corner radius, 5 Draught, 6 Entry radius, 7 Bottom radius.

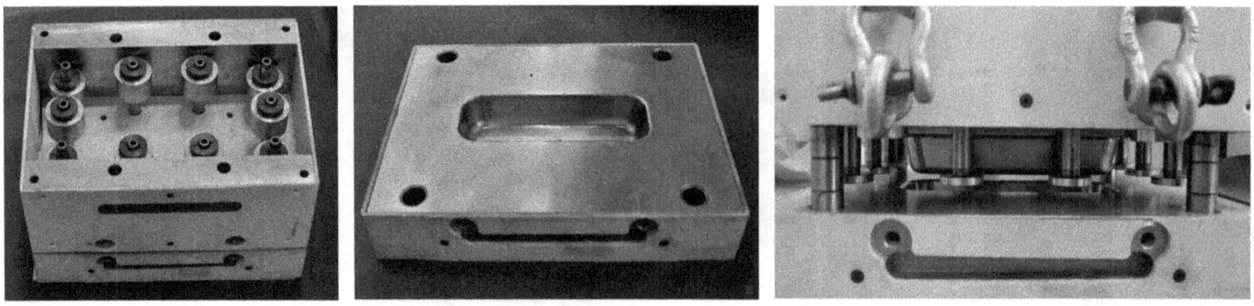

a) b) c)

Figure 4. Validation tool: (a) Top view of the closed tool with binder segment weights, (b) Top view of the opened tool (die only), (c) Side view of the partially closed tool (die, punch, binder segments, guides).

brid yarn. The structure of a multi-layered flat knitted textile is displayed in Fig. 5.

The software deployed for the forming simulation is the FE package PAM-FORM®. In order to gain material properties as input data for the material model, quasi-static material tests are conducted under process temperature. Uniaxial tensile tests, shear tests, bending tests and friction tests are carried out using a universal testing machine equipped with a circulating air oven. The experimental techniques used for shear and bending tests are well established within multiple research groups as the Picture Frame Test (PFT) and the Cantilever Bending Test, respectively (Boisse et al., 2011). Since most of the tests are non-standardised, several custom-built test rigs are utilised. The coefficient of friction is determined for only one fibre direction and the textile/tool friction. The simulation is deemed to be fairly insensitive in terms of the inter-ply coefficient of friction due to the small relative movement. Therefore, the coefficient is estimated to reduce the experimental effort. Tensile and bending tests are performed for warp and weft direction. Fig. 6 gives an overview of the experimental setups.

To verify the quality of the simulation, the material tests are replicated in the simulation software. Calibration of the input data is performed if simulation results differ from the experimental results, thus improving the simulation results of the subsequent FE forming simulations. No calibration of the input data is required apart from the bending module which is reduced iteratively in order to achieve the deflection occurring in the experiments. Fig. 7 shows experimental and simulated curves of the material behaviour.

Due to the architecture of the multi-layered flat knitted textile, each ply has two initial fibre directions (e.g. 0/90°). The forming simulations are performed for two different lay-ups: two plies, both with initial fibre angle $\alpha = 0/90°$ and two plies both with initial fibre angle $\alpha = \pm 45°$. The simulation model of the validation process consists of a simplified punch, die, segmented binder and the composite plies as shown in Fig. 8. The tool parts are modelled as rigid shells with load and displacement constraints. Inter-ply contacts and contacts between textile plies and tool are defined. To represent the textile material properties the implemented composite material model is used.

5. Validation experiments

Validation experiments are conducted under a (nearly) isothermal process regime: two layers of thermoplastic textile are stacked onto the die. The tool is heated to process temperature (200°C). The temperature-controlled infrared panel is used to heat up the composite material from the top. As the material reaches process temperature, the panel is removed and the tool is closed instantaneously. Temperature and pressure are sustained for a defined duration before the temperature is lowered to ejection temperature and the composite part is removed manually.

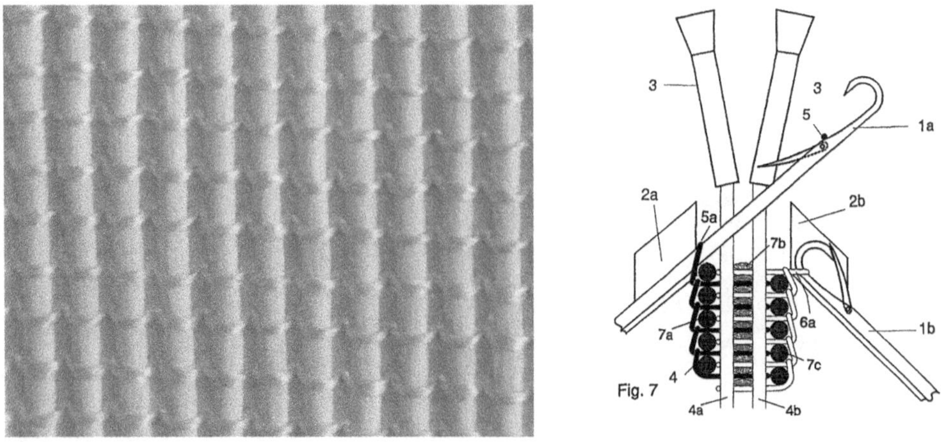

a) b)

Figure 5. Multi-layered flat knitted textile: (a) Left side of the textile in detail, (b) Structure (Engelmann et al., 2002): 4 Warp thread, 7 Weft thread, 5 Stitch thread, 1 Latch needle.

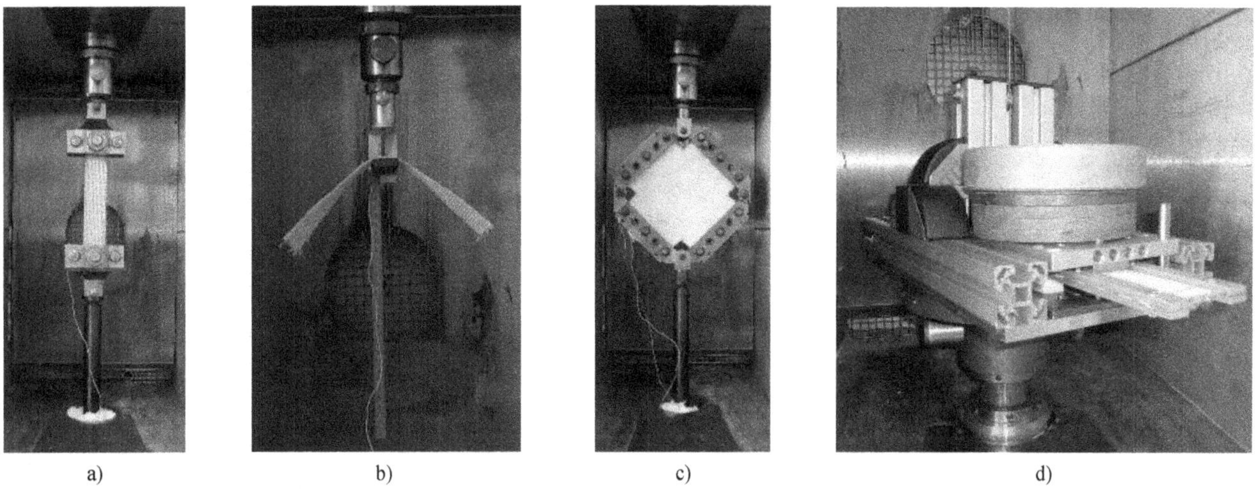

Figure 6. Experimental setups for the material characterisation: (a) Tensile test, (b) Bending test, (c) Shear test (PFT), (d) Friction test.

Validation parts are produced for both lay-ups ($\alpha = 0/90°$ and $\alpha = \pm 45°$). Components made during the experiments are shown in Fig. 9.

6. Numerical and experimental results

As an example, first results produced with the described process are shown in Fig. 10. While the results of draw-in and wrinkling in the draw-in section show good agreement between simulations and experiments, there is a substantial difference between predicted (simulated) and experimental shear angles. Wrinkling in the main component geometry that occurred in the simulation could not be detected in the produced parts.

7. Conclusion

Numerical analyses of forming processes of textile thermoplastics can reduce or replace expensive and time consuming manufacturing studies during the optimisation of

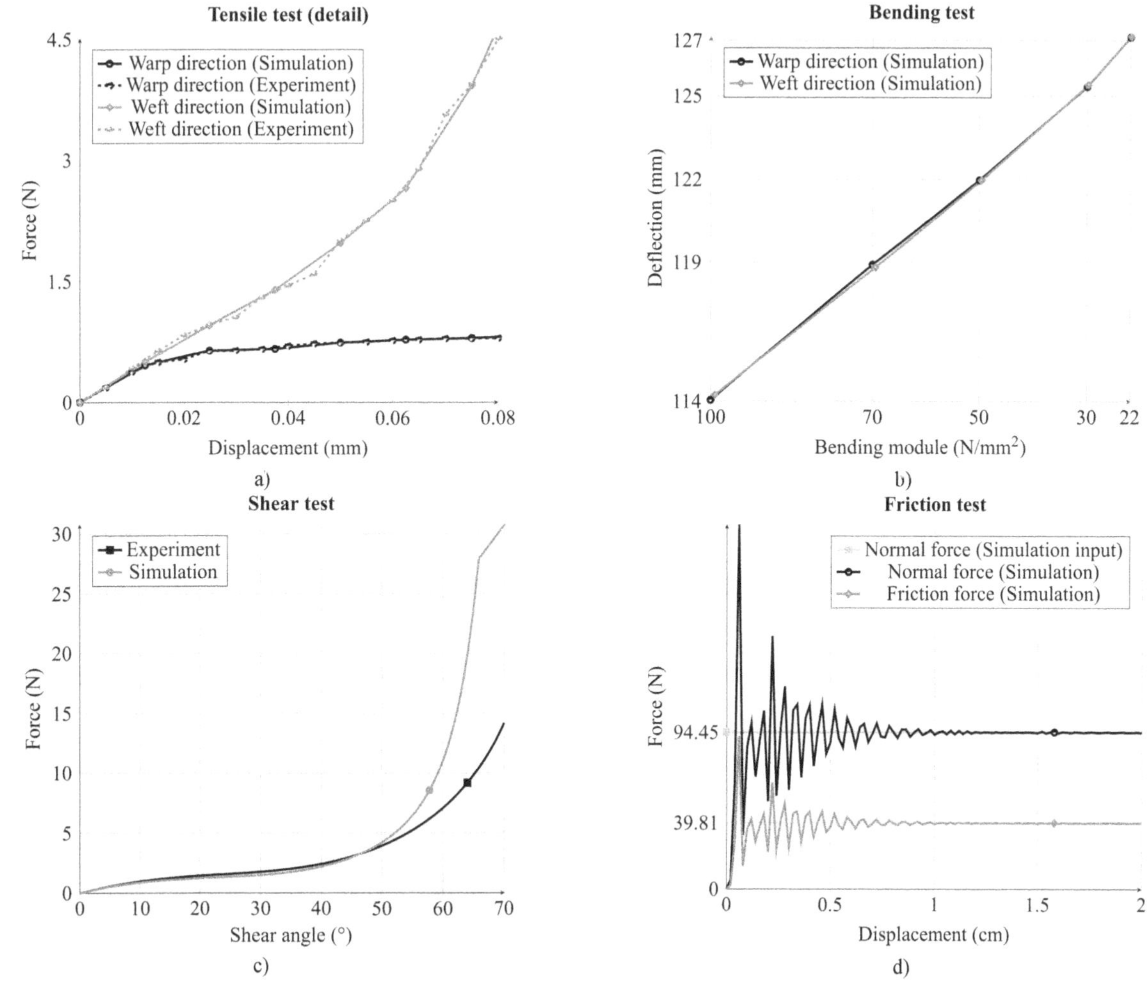

Figure 7. Experimental and simulated curves of the material characterisation: (a) Tensile test, (b) Bending test, (c) Shear test (PFT), (d) Friction test.

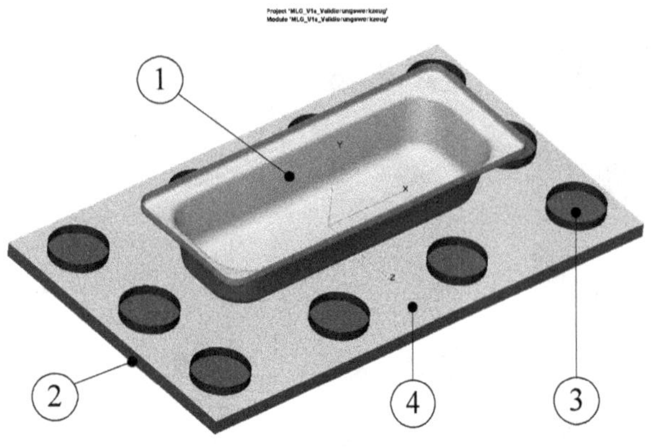

Figure 8. Simulation setup: 1 Punch, 2 Die, 3 Binder segments, 4 Plies.

component geometries and manufacturing processes. Especially the prediction of induced shear angles and wrinkles and the definition of pre-cuts are of great benefit. The focus of this research was the development and trial of a generic thermoforming process for the validation of FE thermoforming simulations. Therefore, typical thermoforming processes were analysed. Kinematic draping simulations were used to define a geometry that causes representative degrees of deformation. The essential material data of the multi-layered flat knitted textile for the FE-based simulation was gained during experimental tests at process temperature. Subsequently, thermoforming simulations were performed. Components produced with the validation tool were compared to the numerical simulation by means of shear angles, draw-in and the occurrence of wrinkles. While the results for draw-in and wrinkling in the draw-in section showed good agreement between simulation and experiment, there were still obvious deviations for shear angles and wrinkling in the main component geometry.

Acknowledgement

The authors gratefully acknowledge the financial support of the Deutsche Forschungsgemeinschaft (German Research Foundation) within the Collaborative Research Center SFB 639, subprojects D4 and E1.

References

Boisse, P., Hamila, N., Vidal-Sallé, E. and Dumont, F. (2011). Simulation of wrinkling during textile composite reinforcement forming. Influence of tensile, in-plane shear and bending stiffnesses. *Composites Science and Technology*, 71*(5)*, pp. 683-692.

Brooks, R. (2007). Forming technology for thermoplastic composites, in Long, A.C. (ed.). *Composites forming technologies*. Cambridge: Woodhead, pp. 256-276.

Cherouat, A. and Borouchaki, H. (2009). Present state of the art of composite fabric forming: geometrical and mechanical approaches, in *Materials, 2(4)*, pp. 1835-1857.

Engelmann, W., Hoffmann, G. and Offermann, P., (Patent of the Technische Universität Dresden), (2002), *Multilayer knitted structure and method of producing the same*. European patent EP0873440 B1. 2002 -04-10.

Hufenbach, W., Langkamp, A., Adam, F., Krahl, M., Hornig, A., Zscheyge, M., Modler, K.H. (2011). An integral design and manufacturing concept for crash resistant textile and long-fibre reinforced polypropylene structural components. *Procedia Engineering 10* (2011), pp. 2086–2091.

Hufenbach, W., Modler, N., Krahl, M., Hornig, A., Ferkel, H., Kurz, H. and Ehleben, M. (2010a). Leichtbausitzschalen im Serientakt. Integrales Bauweisenkonzept, in *Kunststoffe 100* (2010), pp. 56-59. (in German)

Hufenbach, W., Meschke, J., Dannemann, M. and Friebe, S. (2010b). Masseneutrale Minderung des Fahrzeuginnengeräusches durch den Einsatz einer neuartigen Reserveradmulde aus textilverstärkten Thermoplastverbunden, in *Proceedings of the 36. Deutsche Jahrestagung für Akustik* – DAGA 2010, Berlin, Germany, March 15 -18, pp. 113-114. (in German)

Hufenbach, W., Dannemann, M., Friebe, S., Kolbe, F. and Krahl, M. (2009). Design and tests of thermoplastic textile-reinforced composite trays for vibro-acoustic relevant applications, in *Proceedings of the International Conference on Acoustics* NAG/DAGA 2009, Rotterdam, The Netherlands, March 23-26.

Khan, M.A., Mabrouki, T., Vidal-Salle, E. and Boisse, P. (2010). Numerical and experimental analyses of woven composite reinforcement forming using a hypoelastic behaviour. Application to the double dome benchmark. *Journal of Materials Processing Technology, 210 (2)*, pp. 378-388.

Lee, J.S., Hong, S.J., Yu, W.R. and Kang, T.J. (2007). The effect of blank holder force on the stamp forming behavior of non-crimp fabric with a chain stitch. *Composites Science and Technology*, 67*(3)*, pp. 357-366.

Nino, G.F., Bersee, H.E.N. and Beukers, A. (2008). Design and Manufacturing of Thermoplastic Composite Ribs Based on Finite Element Analysis, in 49th AIAA/ASME/ASCE/AHS/ASC Structures, Structural Dynamics, and Materials Conference, Schaumburg, USA, April 07-10.

Pantelakis, S.G., Katsiropoulos, C.V., Labeas, G.N. and Sibois, H. (2009). A concept to optimize quality and cost in thermoplastic composite components applied to the production of helicopter canopies. *Composites Part A: Applied Science and Manufacturing*, 40*(5)*, pp. 595-606.

Van Der Weeën, F. (1991). Algorithms for draping fabrics on doubly curved surfaces. *International Journal for Numerical Methods in Engineering*, 31*(7)*, pp. 1415–1426.

Willems, A., Lomov, S.V., Vandepitte, D. and Verpoest, I. (2006). Double dome forming simulation of woven textile composites, in *Proceedings of the 9th ESAFORM conference on material forming*. Publishing House Akapit, Poland, pp. 747-750.

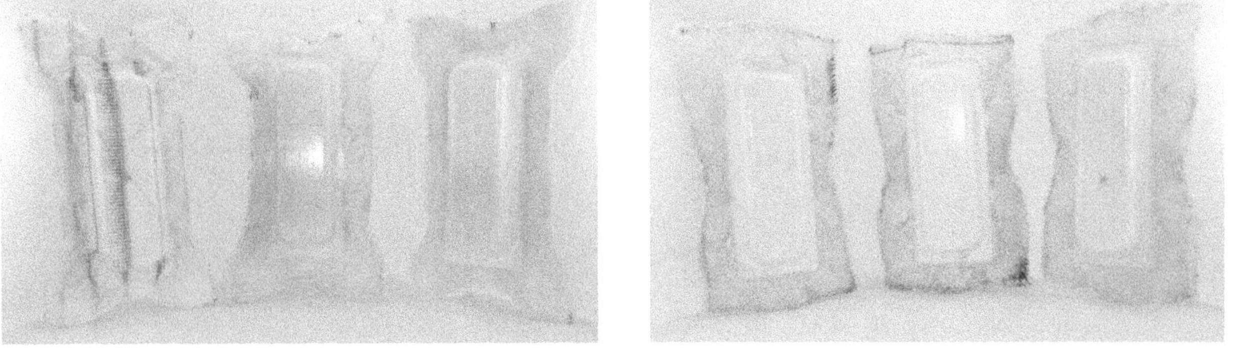

Fibre angle $\alpha = 0/90°$ Fibre angle $\alpha = \pm 45°$

Figure 9. Validation components.

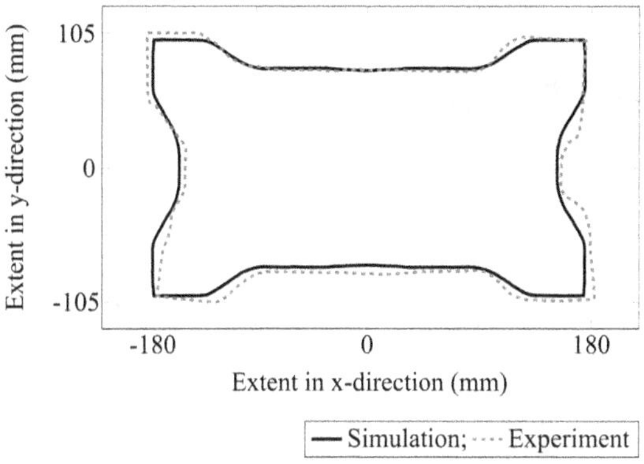

Fibre angle α = 0/90°

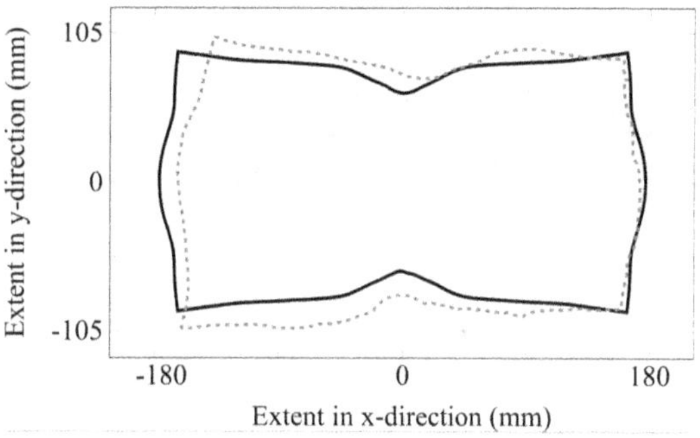

Fibre angle α = ±45°

a)

Experiment, Fibre angle α = 0/90°

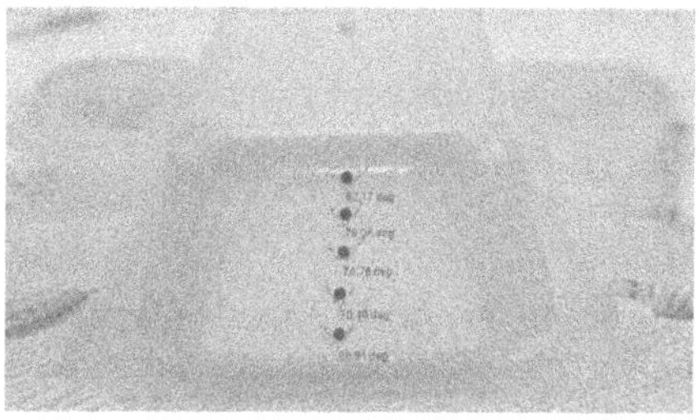

Experiment, Fibre angle α = ±45°

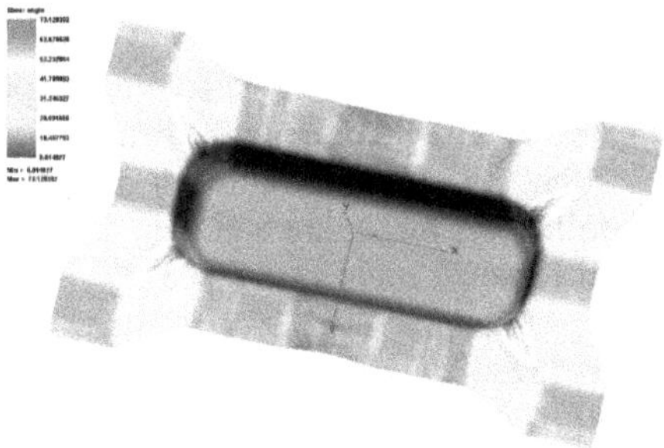

Simulation, Fibre angle α = 0/90°

Simulation, Fibre angle α = ±45°

b)

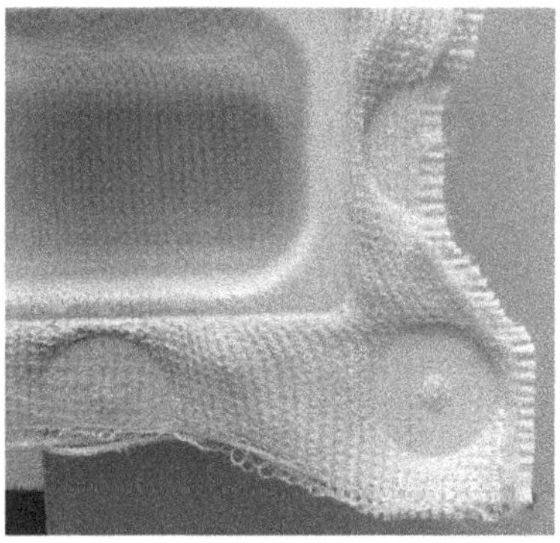

Experiment, Fibre angle α = 0/90°

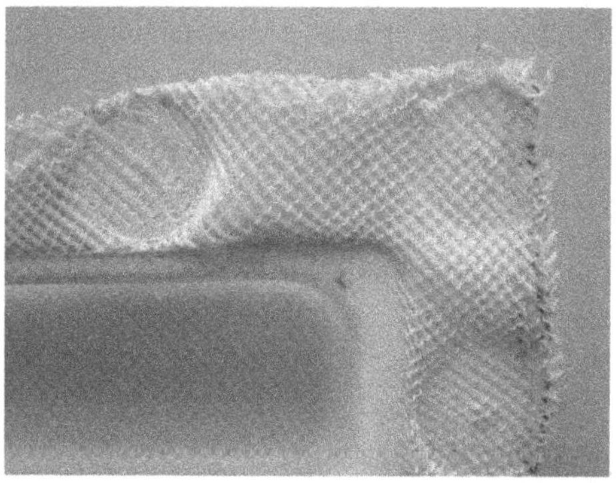

Experiment, Fibre angle α = ±45°

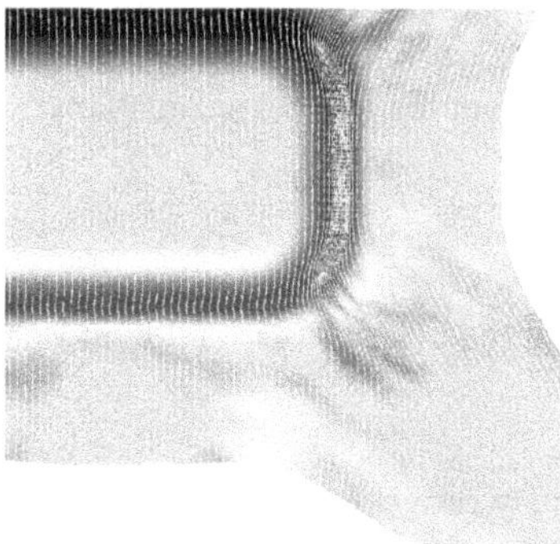

Simulation, Fibre angle α = 0/90°

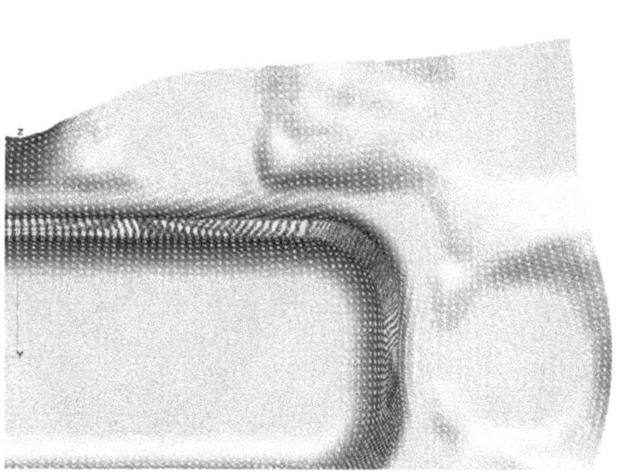

Simulation, Fibre angle α = ±45°

c)

Figure 10. Experimental results and simulation results for the initial fibre angles α = 0/90° and α = ±45°: (a) Draw-in (upper ply), (b) Shear angles (lower ply), (c) Wrinkles (upper ply).

Design of Anisogrid Composite Lattice Beams for Spacecraft Applications

Andrey Azarov

Rocket and Spacecraft Composite Structures Department, Faculty of Special Machinery, Bauman Moscow State Technical University, 5 2nd Baumanskaya Street, Moscow, 105005, Russia

Abstract: The paper is concerned with analysis and design of Anisogrid (Anisotropic Grid) thin-walled beams consisting of a system of helical and hoop ribs made of unidirectional carbon-epoxy composite material by continuous filament winding. Such beams are used in spacecraft structures including trusses, solar panels and space antennas and must have high stiffness and minimum mass. The design of Anisogrid thin-walled beam of minimum mass under stiffness constraints is considered.

Key Words: Anisogrid structures, Composite materials, Thin-walled beams.

1. Introduction

Thin-walled beams are widely used in spacecraft structures including trusses, solar panels and space antennas. These structures do not experience the action of considerable loads in the process of launching and operation in space, must have high stiffness and, what is most important, minimum mass. The most promising material for such elements is carbon-fiber reinforced composite having high specific (with respect to density) strength and stiffness, and low thermal expansion coefficient. There are two possible structural forms of composite thin walled beam – laminated beam, consisting of a number of composite plies (Fig. 1a) and Anisogrid lattice beam (Fig. 1b).

Anisogrid (Anisotropic Grid) structures were proposed about thirty years ago and are under serial production in Central Research Institute of Special Machinery (CRISM) which develops lattice interstages and payload attach fittings (adapters) for Russian heavy launcher Proton-M developed by Khrunichev Space Center and having by now about 55 successful launches with lattice structures (Vasiliev et al., 2012).

In contrast to smooth thin-walled laminated beams consisting of a number of composite plies with finite thickness whose mass cannot be reduced beyond some minimum value determined by the thickness of the ply, the mass of Anisogrid lattice beams is governed by the rib spacing and cross-sectional area and can be readily reduced to a desirable value. The minimum unit length mass of the existing actual lattice beams reaches 0.25 kg/m.

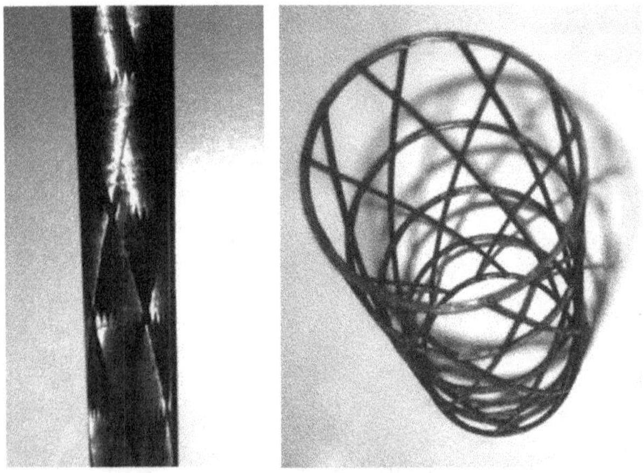

a) b)

Figure 1.Thin-walled composite beams: laminated beam (a) and Anisogrid beam (b).

2. Governing equations

Consider a thin-walled beam shown on Fig. 2. The axial coordinate z is measured along the beam axis and the contour coordinate s is measured from a certain point A (Fig. 2). The following forces are acting in the beam cross section: axial force N, transverse forces Q_x, Q_y, bending moments M_x, M_y, and torque H. Axial stress resultant n and shear stress resultant q, acting in the wall are determined by the following equations (Vasiliev, 1993)

$$n = B \cdot \left(\frac{N}{S} + \frac{M_x}{D_x} y + \frac{M_y}{D_y} x \right) \tag{1}$$

$$q = Q_x F_x(s) + Q_y F_y(s) + \frac{H}{2F} \tag{2}$$

where

$$S = \oint B ds \; ; \; D_x = \oint B y^2 ds \; ; \; D_y = \oint B x^2 ds \tag{3}$$

- axial and bending stiffnesses with respect to coordinate axes x and y, B– wall stiffness coefficient;

$$F_x(s) = -\frac{1}{D_y} \int_0^s Bxdx \quad ; \quad F_y(s) = -\frac{1}{D_x} \int_0^s Bydx \tag{4}$$

and F is the area bounded by the crossectional contour.

Axial deformation and the beam cross-section angles of rotation at the direction of bending moments M_x and M_y can be found from the following equations:

$$u'_z = \frac{N}{S}; \; \theta'_x = \frac{M_x}{D_x}; \; \theta'_y = \frac{M_y}{D_y} \tag{5}$$

where $( )' = d( )/dz$, u_z – axial displacement. For the displacements of the beam axis u_x and u_y in x and y directions we have

$$u'_x = \psi_x - \theta_y; \; u'_y = \psi_y - \theta_x; \tag{6}$$

where y_x and y_y are the shear deformations in xz and yz planes, respectively, and are determined as

$$\psi_x = \frac{Q_x}{K_x} \qquad \psi_y = \frac{Q_y}{K_y} \tag{7}$$

in which

$$K_x = \left(\oint F_x^2(s) \frac{dS}{C} \right)^{-1} \quad K_y = \left(\oint F_y^2(s) \frac{dS}{C} \right)^{-1} \tag{8}$$

are the beam shear stiffnesses. Functions $F(s)$ are determined by Eq. (4), C is the wall shear stiffness coefficient. Finally, for the twist angle q we have

$$\theta = \frac{H}{4F^2} \oint \frac{dS}{C} \tag{9}$$

The forces and moments acting in the beam cross section (Fig. 2) can be found using the equilibrium equations

$$\begin{aligned} &N' + f_z = 0; \\ &Q_y' + f_y = 0; \\ &M_x' - Q_y + m_x = 0; \\ &M_y' - Q_x + m_y = 0; \\ &H' + m_z = 0; \end{aligned} \tag{10}$$

where f and m are distributed forces and moments acting on the beam. Equations (10) have a sixth order and the solution contains three forces and three moments acting on one end of the beam, for example, $z = 0$.

Integration of Eqs. (5), (6) and (9) leads to a solution, which includes three displacements, and three rotation angles for $z = 0$ cross section. The general solution contains 12 integration constants, which can be found using the boundary conditions at the ends of the beam. Six boundary conditions are formulated at each end. At the free end three forces and three moments are zero, at the simply supported end $M_x = M_y = H = 0$, $N = 0$ and $u_x = u_y = 0$ and at the fixed end three displacements and three rotation angles are zero.

The foregoing equations include two wall stiffness coefficients B and C. For the thin-walled beam without transverse walls we have (Vasilvev, 1993)

$$B = B_{11} - \frac{B_{12}^2}{B_{22}} \quad C = B_{33} \tag{11}$$

For the Anisogrid beam, consisting of a system of unidirectional composite helical, and circumferential ribs, stiffness coefficients in Eq. (11) can be found as

$$\begin{aligned} &B_{11} = 2\frac{B_h}{a_h}\cos^4\varphi \\ &B_{12} = B_{33} = 2\frac{B_h}{a_h}\cos^2\varphi\sin^2\varphi \\ &B_{22} = 2\frac{B_h}{a_h}\sin^4\varphi + \frac{B_c}{a_c} \end{aligned} \tag{12}$$

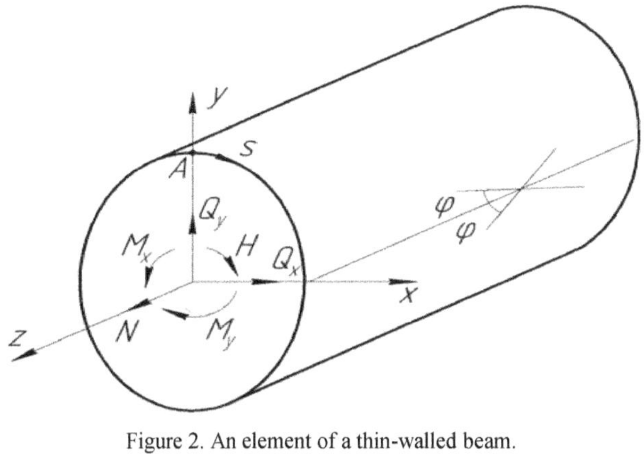

Figure 2. An element of a thin-walled beam.

where

$$B_h = E_h h \delta_h \quad B_c = E_c h \delta_c \tag{13}$$

are ribs axial stiffnesses. In Eqs. (12), (13), subscript "h" corresponds to helical ribs, subscript "c" – to circumferential ribs, φ is the helical ribs angle, a is the rib spacing (counted along the normal to the rib axes), E is the rib modulus (in general, different for helical and circumferential ribs), δ – rib widths, h – rib height. For the ribs spacings the following relationships are valid:

$$a = \frac{2\pi R}{n} \quad a_h = a\cos\varphi \quad a_c = \frac{a}{2\tan\varphi} \tag{14}$$

where n is the number of symmetric pairs of helical ribs. The mass of lattice structure per unit area is determined as

$$m = \left(2\rho_c \frac{\delta_c}{a_c} + \rho_\kappa \frac{\delta_\kappa}{a_\kappa} \right) h \tag{15}$$

where ρ is the rib density.

Consider the beam with circular cross-section under three-point bending. Both ends of the beam are simply supported and transverse force P is applied to the middle of the beam. The length of the beam is l. For such a beam $f = m = 0$ in Eq. (10). Integrating Eq. (10) with the boundary conditions $Q_y(z{=}0) = P/2$, $M_x(z{=}0) = 0$, we have

$$Q_y = \frac{P}{2}; \; M_x = \frac{Pz}{2}$$

For the circular cross section we have $y = R\cos\bar{s}$ where $\bar{s} = s/R$ (Fig. 2) and the beam bending and shear stiffnesses can be found in accordance with Eqs. (3), (4) and (8):

$$D_x = \pi B R^3 \quad F_y(s) = \frac{BR^2}{D_x}\sin\bar{s} \quad K_y = \pi R C \tag{16}$$

Integrating second Eq. (5) with regard to Eq. (15) and the boundary condition $q_x(z{=}l/2){=}0$, we obtain the cross section rotation angle as

$$\theta_x = \frac{P}{16D_x}\left(4z^2 - l^2\right) \tag{17}$$

Substituting y_y and q_x from Eqs. (7) and (17) to Eq. (6) and integrating with condition $u_y(z = 0) = 0$, we arrive at the following equation for the beam displacements along y axis

$$u_y = \frac{Pz}{16D}\left(3l^2 - 4z^2 + 24\frac{D_x}{K_y} \right)$$

For the maximal displacement $v = u_y(z = l/2)$, we have

$$v = \frac{Pl^3}{48D_x}(1 + \lambda) \tag{18}$$

where

$$\lambda = \frac{12D_x}{K_y l^2} = 12\frac{BR^2}{Cl} \tag{19}$$

is the coefficient determining the deflection due to shear deformation.

3. Axial stiffness of the beam

Consider Anisogrid beam like shown on Fig. 1b with circular cross section (R = 54.5 mm) and length l=6060 mm. The beam consists of a system of helical and circumferential ribs with the following parameters:

- rib height $h = 2.85$ mm;
- rib width $\delta_h = \delta_c = 2,85$ mm;
- helical rib angle $\varphi = 15°$;
- helical rib spacing $a_h = 56.6$ mm;
- circumferential ribs spacing, $a_c = 109.3$ mm;
- rib modulus $E_h = E_c = E = 180$ GPa;
- rib density $\rho_h = \rho_c = \rho = 1462$ kg/m^3.

According to Eq. (15), the structure has mass per unit area $m = 0,525$ kg/m^2. Stiffness coefficients can be found using Eqs. (11), (12) and (13), which give

$$\begin{aligned} &B_h = B_c = 1460 \text{ GPa}\times\text{mm}^2;\\ &B_{12} = 44.63 \text{ GPa}\times\text{mm};\\ &B_{12} = B_{33} = 3.21 \text{ GPa}\times\text{mm};\\ &B_{22} = 13.57 \text{ GPa}\times\text{mm};\\ &B = 43.87 \text{ GPa}\times\text{mm};\\ &C = 3.21 \text{ GPa}\times\text{mm}. \end{aligned} \tag{20}$$

The maximal beam displacement under three-point bending with force P = 20 N is determined by (18), according to which $v = 4.2$ mm. However, experimental value for such beam was found to be $v = 7.9$ mm, which is 1.9 times higher than the calculated value.

To determine the effect of the actual stiffness reduction, the element of a beam consisting of nine sections, was subjected to axial compression (Fig. 3a). The axial displacement under the force 7 kN was found to be 1.08 mm, which corresponds to the axial wall stiffness coefficient B of 19.7 GPa×mm, which is 2.2 times lower than the value specified by Eq. (20). The finite element analysis have also been performed. The ribs were modeled by beam elements with approximate number of ten elements between ribs intersections. The deformed state of finite element model is shown in Fig. 3b. As can be seen in Fig. 3b, the helical ribs segments between circumferential ribs experience bending reducing the axial stiffness of the structure.

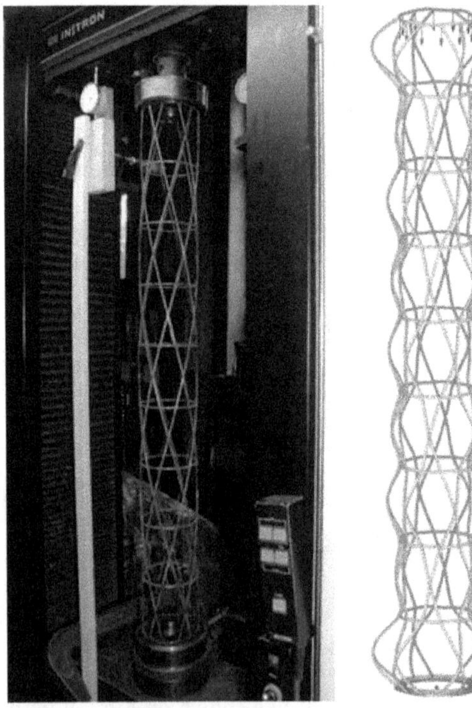

a) b)

Figure 3. The element of Anisogrid beam under axial compression (a) and the corresponding finite element model (b).

To take this effect into account helical rib segment must be considered as a compressed curved beam, which leads to the following equation for the rib stiffness

$$B_h^t = \frac{B_h}{1+\eta} \tag{21}$$

where B_h is determined by Eq. (13) and

$$\eta = \frac{B_h t_0^2}{D_h}\left(\frac{1}{2} - \frac{4}{\pi^2}\right) \tag{22}$$

is the coefficient determining the rib stiffness reduction due to bending. In this equation t_0 depends on the beam structural parameters

$$t_0 = \frac{R}{\sin^2\varphi}\left[1 - \cos\left(\frac{a_c \sin^2\varphi}{2R\cos\varphi}\right)\right]$$

where $D_h = E_h h^3 \delta_h/12$ is the rib bending stiffness.

For the considered beam we have $t_0 = 1.97$ mm, $B_h = 1460$ GPa×mm^2 and $D_h = 990$ GPa×mm^4 and (22) gives $\eta = 0.54$. Thus, $B^t_h = 948$ GPa×mm^2. Substituting B^t_h instead of B_h in Eq. (12) and using Eq. (18) to find the maximum displacement, we arrive at $v = 6.5$ mm, which is 18% lower than the experimental value 7.9 mm

Thus, the displacement of the Anisogrid beam under three-point bending is determined by Eq. (18) and the stiffness coefficients Eq. (12) must be found using the reduced rib stiffness Eq. (21).

4. Design

Consider the design of the Anisogrid beam of minimum mass under stiffness constraints. The design parameters are φ, h, δ_h, δ_c, a_h and a_c. The objective function is the beam mass, which can be found by multiplying the mass per unit area from Eq. (15) on the surface area. For the beam with circular cross section, we have

$$M = 2\pi R l\left(2\rho_h \bar{\delta}_h h + \rho_c \bar{\delta}_c h\right) \tag{23}$$

where $\bar{\delta}_h = \frac{\delta_h}{a_h}$; $\bar{\delta}_c = \frac{\delta_c}{a_c}$

The stiffness constraint has the form of restriction imposed on the maximal deflection of the beam under three-point bending $\bar{v}$. Using Eq. (18) we obtain

$$\bar{v} = \frac{Pl^3}{48D_x}(1+\lambda) \tag{22}$$

To solve the design problem, we introduce some simplifications. In the Eq. (2) for the wall stiffness coefficient B, we neglect second term B^2_{12}/B_{22}, which is proportional to $\sin^4\varphi$ according to (12). In actual beams the angle φ is small enough and this term is considerably less than B_{11}, which is proportional to $\cos^4\varphi$. The resulting axial stiffness B corresponds to the beam with undeformable crossectional contour. This simplification reduces axial, shear and bending stiffnesses described by Eqs. (11) and (16) to the form

$$\begin{aligned} B &= 2\frac{B_h^t}{a_h}\cos^4\varphi \\ C &= 2\frac{B_h^t}{a_h}\cos^2\varphi\sin^2\varphi \\ D_x &= 2\pi R^3\frac{B_h^t}{a_h}\cos^4\varphi \end{aligned} \tag{25}$$

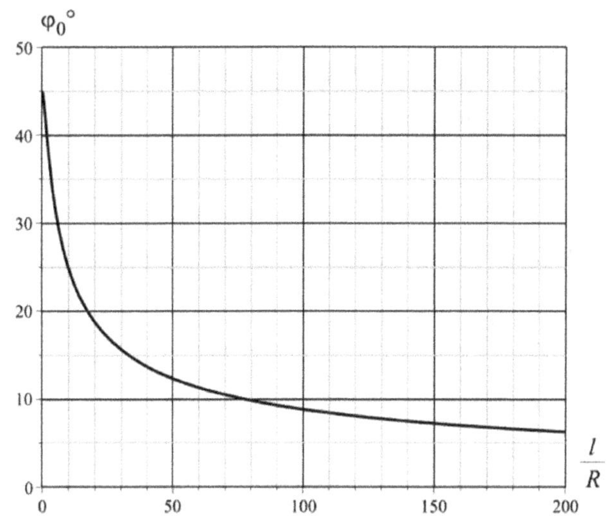

Figure 4. The dependence of the optimal angle of helical ribs on the normalized length of the beam.

where B^t_h is determined by Eq. (21). Due to the complexity of the B^t_h coefficient that depends on design parameters according to Eq. (22), we allow for the local ribs bending using an iterative process involving successive refinement of the η parameter.

As a first approximation we assume η = 0. Equations (25) take the form

$$B = 2E_h h\bar{\delta}_c c^4$$
$$C = 2E_h h\bar{\delta}_h c^2(1-c^2) \quad (26)$$
$$D_x = 2\pi R^3 E_h h\bar{\delta}_h c^4$$

where $c = \cos\varphi$. Substituting this coefficients into Eq. (24), we obtain

$$\bar{v} = \frac{Pl^3}{96\pi E_h h\bar{\delta}_h R^3}\left(\frac{1}{c^4} + 12\frac{R^2}{l^2}\frac{1}{c^2(1-c^2)}\right) \quad (27)$$

Using this equation we can find helical ribs normalized area $\bar{\delta}_c h$ as

$$\bar{A}_c = \bar{\delta}_c h = \frac{Pl^3}{96\pi E_c R^3 \bar{v}}\left(\frac{1}{c^4} + 12\frac{R^2}{l^2}\frac{1}{c^2(1-c^2)}\right) \quad (28)$$

Substituting Eq. (28) into Eq. (23) and using the condition for the minimum $dM/dc = 0$ of the beam mass, we can obtain the c value corresponding to the minimum mass of the beam, i.e.,

$$c = \sqrt{\frac{3\bar{R}^2 + \sqrt{6\bar{R}^2 + 9\bar{R}^4} - 1}{12\bar{R}^2 - 1}}$$

where $\bar{R} = R/l$. Since $c = \cos\varphi$, the optimal angle φ_0 becomes as

$$\varphi_0 = \arccos\left(\sqrt{\frac{3\bar{R}^2 + \sqrt{6\bar{R}^2 + 9\bar{R}^4} - 1}{12\bar{R}^2 - 1}}\right) \quad (29)$$

The dependence of the optimal angle of helical ribs on the normalized length of the beam l/R is shown on Fig. 4.

Knowing the optimal angle from Eq. (29) and using Eq. (28) we can find the normalized cross-sectional area of helical ribs $\bar{A}_h = \bar{\delta}_h h$. Note that the applied design model allows us to obtain only normalized area of the rib, the dimensions of the rib follow from applied manufacturing considerations, as well as the number of helical ribs or spacings. It is advisable to choose firstly the number of pairs of helical ribs n, passing through the cross section. Then, using Eq. (14) with $\varphi = \varphi_0$, the helical ribs spacing a_h can be found. Assuming further that the helical rib has a square or rectangular cross section and determining the ratio δ_h/h, the rib width d_h and height h which provide the required rib cross section area $A_h = a_h\bar{A}_h = \delta_h h$ can be found. The width of circumferential ribs must take the minimal value acceptable by manufacturing conditions.

In the second approximation step the local bending of helical ribs is considered.The parameter h in Eq. (6) for the stiffness of the helical ribs, is specified by Eq. (7) and Eq. (8) for the design parameters obtained in the first approximation. On each iteration the new h value is calculated. On the i-th iteration, the stiffness coefficients have the following form

$$B_i = \frac{2E_h h_h \bar{\delta}_h c^4}{1+\eta_{i-1}}$$
$$C_i = \frac{2E_h h_i \bar{\delta}_{hi} c^2(1-c^2)}{1+\eta_{i-1}} \quad (30)$$
$$D_i = \frac{2\pi R^3 E_h h_i \bar{\delta}_{hi} c^4}{1+\eta_{i-1}}$$

Substituting Eq. (30) in Eq. (24), we obtain the displacement for the iteration number i

$$\bar{v} = (1+\eta_{i-1})\frac{Pl^3}{96\pi E_c h_i \bar{\delta}_{ci} R^3}\left(\frac{1}{c^4} + 12\frac{R^2}{l^2}\frac{1}{c^2(1-c^2)}\right)$$

Solving this equation for the relative helical rib area, we have

$$\bar{\delta}_{hi} h_i = \frac{(1+\eta_{i-1})Pl^3}{96\pi E_h R^3 \bar{v}}\left(\frac{1}{c^4} + 12\frac{R^2}{l^2}\frac{1}{c^2(1-c^2)}\right)$$

Then, the new values h_i, $\bar{\delta}_{hi}$, A_{hi} and η_i can be found. The difference between η_{i-1} and η_i is calculated in every step. If the difference is less than the certain value, the process ends. The results obtained in the last iteration are the final optimal parameters of the lattice structure.

Conclusion

The bending of the helical ribs segments between circumferential ribs, which reduce the axial stiffness of the structure should be taken into account during the analysis and design of Anisogrid lattice beams in order to obtain more accurate results.

References

Vasiliev, V.V. (1993).*Mechanics of Composite Structures*. Washington: Tailor & Francis.

Vasiliev, V.V. and Razin, A.F. (2006). Anisogrid composite lattice strucrures for spacecraft and sircraft applications, *Composite Structures*, Vol. 76, pp.182-189.

Vasiliev, V.V., Barynin, V.A. and Razin, A.F. (2012). Anisogrid composite lattice structures – Development and aerospace applications, *Composite Structures*, Vol. 94, pp.1117-1127.

Vasiliev, V.V., Barynin, V.A. and Razin, A.F. (2001) Anisogrid lattice structures – Survey of development and application, *Composite Structures*, Vol. 54, pp. 361-370.

Vasiliev, V.V.and Razin, A.F. (2001) Optimal design of filament – Wound anisogrid composite lattice structures, *Proc. Of the 16-th Annual Techn. Conf. of American Society for Composites*, Blacksburg, USA, (CD).

Totaro, G., Vasiliev, V.V. and De Nicola, F. (2004) Optimized design of isogrid and anisogrid lattice structures. *Proc. Of the 55-th int. Astronautical Congr*, Vancouver, Canada, (CD).

Multi-criteria Design Pressure Vessel Manufactured from Composite Materials by the Method of Winding

Sergey Gavriushin, Minh Dang

Computer Systems of Production Automation Department, Faculty of Robotics and Complex Automation, Bauman Moscow State Technical University,5 2nd Baumanskaya Street, Moscow, 105005, Russia

Abstract: This article describes the unified methodology of design of a composite pressure vessel by winding in the concept of life-cycle management products. The proposed algorithm is based on a multi-criteria optimization and the parameter space investigation (PSI) method for the decision of tasks of synthesis. In the work it is given as an example of the process of winding along a geodesic path.

Key Words: Automation, Product life-cycle, Computer-aided design, CAD, Optimization, Multi-criteria, Design, Composite material, Winding, Pressure vessel, PSI.

The process of continuous winding is one of the most common and committed processes of the production of high-strength composite shells, which are widely used in aviation, automobile industry and other branches of industry.

Production of pressure vessels made of composite materials, in modern conditions, requires the use of the concept product life-cycle management (PLM) in a unified information space (UIS) (Dang and Gavriushin, 2012). Currently, engineers, technologists and manufacturers have gained large experience in the field of design, calculation and manufacture of composite products (Bratukhin et al., 2003; Vasiliev, 1988; Bulanov et al., 1985; Komkov and Tarasov, 2011; Sarbaev, 2001). However, practice shows that at the junction of the individual phases of the life-cycle (formulation of initial requirements in design, strength analysis, production technology, etc.), there remain elements of inconsistency, and a pair, and even contradictions that hamper the process of production as a whole. For example, calculations of the designer may be unrealizable for the chosen technology, or better technology may not be effective for economic or other reasons. Therefore, in order to provide a consistent mechanism for collaboration among participants in the life-cycle, it is necessary to solve the problem of synthesising rational production options (Fig. 1).

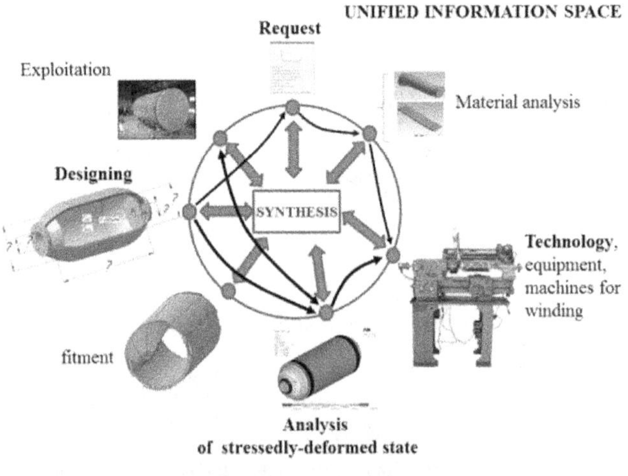

Figure 1. The synthesis problem in a simplified life-cycle of products made from composite material by the method of winding.

In general, the problem of synthesis of the winding process of composite pressure vessel within the concept of the PLM is multi-criteria. To satisfy the requirements of, (or improving the criteria of), some of the participants in the life-cycle, for example, calculators, designers, etc. we, as a rule, may not meet the requirements of, (or impair the criteria of), other participants (technologists, manufacturers, etc.). In addition, since the cylinder pressure from composite materials is a high-tech product (Sarbaev, 2001), the customer may and correctly formulate the initial data, which complicates the manufacturing process. In the worst case, we may get production solutions of incorrect requirements.

In the present work, we proposed the multi-criteria approach for solving the problem of synthesis of process winding a composite pressure vessel. This approach makes it possible to take into account the requirements, parameters, functional and criterial constraints of all participants in the product life-cycle. The technique is based on the parameter space investigation method (PSI method) (Statnikov et al., 2012). In accordance with this method, to find the optimal parameters it is necessary, first of all, to build the so-called region of admissible solutions, determined by restrictions on the criteria, parameters and functional dependencies. The basis of this region is determined by the set of Pareto-optimal variants, which are characterized by the fact that any of the solutions belonging to this set, we can't improve on all criteria simultaneously.

The given mathematical model of the cylinder and the wishes of various participants in the life-cycle of products that define the range of the criteria of product's quality is; $\overline{\Phi}(\overline{\alpha}) = \left(\Phi_\nu^* \le \Phi_\nu(\overline{\alpha}) \le \Phi_\nu^{**}\right); \nu = \overline{1,M}$. It is necessary to find the acceptable design and technological solutions $\overline{\Phi}\left(\overline{\alpha}^{(i)}\right) = \left(\Phi_1\left(\overline{\alpha}^{(i)}\right),\dots,\Phi_M\left(\overline{\alpha}^{(i)}\right)\right)$, belonging to the region $\underline{\overline{\Phi}}$. Addressing the defined Pareto-optimal solutions selects the most preferred solution $\overline{\Phi}\left(\overline{\alpha}^{\oplus}\right)$. Algorithm of multi-criteria synthesis of the winding process, based on the PSI method, is shown below.

<u>Step 1</u>: Participants in the product life-cycle are determined:

- the boundaries of the change of control parameters (Π_0 parallelepiped in *N*-dimensional space of parameters)

$\alpha_j^* \le \alpha_j \le \alpha_j^{**}; j = \overline{1, N}$.

- the functional constraints $\overline{f}(\overline{\alpha}) = f_1(\overline{\alpha}), f_2(\overline{\alpha}),..., f_K(\overline{\alpha})$. Functional constraints are written in the form of mathematical and (or) logical equations and inequalities.

- the vector of performance criterions $\overline{\Phi}(\overline{\alpha}) = \Phi_1(\overline{\alpha}), \Phi_2(\overline{\alpha}),..., \Phi_M(\overline{\alpha})$ and the area of target values of criteria $\overline{\Phi}$.

Step 2: In a parallelepiped Π_0 by the $\mathbf{LP}_\tau$ – sequence (Sobol and Statnikov, 2006) $\overline{\alpha}^{(l)}; l = \overline{1, N_0}$ generated N_0 test vectors and implemented calculation and verification $\overline{f}\left(\overline{\alpha}^{(l)}\right); l = \overline{1, N_0}$ of the functional constraints.

Step 3: With M_0 vectors ($M_0 \le N_0$), which satisfy constraints, vectors of criteria $\overline{\Phi}\left(\overline{\alpha}^{(i)}\right); i = \overline{1, M_0}$ are calculated and table tests are made. In the table tests the values of criteria are ranked in order of deterioration. Using the table of tests, are appointed by the acceptable constraints for each criteria $\Phi_\nu\left(\overline{\alpha}^{(i)}\right) \le \Phi_\nu^{**(0)}; i = \overline{1, M_0}; \nu = \overline{1, M}$. Get P_0 vectors ($P_0 \le M_0$), satisfying all the constraints.

Step 4: In the space of criteria P_0 acceptable solutions correspond to the P_0 points in the parameter space (and vice versa). A range of change of parameters is divided into intervals, each of which is placed in correspondence with the number of admissible solutions belonging to this interval. On the basis of this analysis are formulated new borders of variations of the parameters for the subsequent correction.

Step 5: The optimization is carried out through the correction of the borders of variations of the parameters, the number of generated sampling points and the tightening of criteria constraints (steps 2-4) until we either get in the area of target values $\overline{\Phi}$, or the maximum approach to the region (k-correction), in case of reaching the target values $\overline{\Phi}$ of the process end.

Otherwise, we need to find the reason for the lack of a solution and to find an acceptable option, changing the setting of the task with the involvement of all participants in the product life-cycle. When searching for an acceptable variant it proves to be useful to use the following visualization tools:

- Graphics «criterion - parameter», which represent the projection of points $\left(\Phi_\nu\left(\overline{\alpha}^{(i)}\right), \alpha_j^{(i)}\right)$; $\nu = \overline{1, M}$; $j = \overline{1, N}$; $i = \overline{1, P_k}$ on plane $\left(\Phi_\nu \alpha_j\right)_k$.
- Graphics «criterion - criterion», comparing the projection of points $\left(\Phi_\nu\left(\overline{\alpha}^{(i)}\right)\right)$; $\nu = \overline{1, M}$; $i = \overline{1, P_k}$ on plane $\left(\Phi_r \Phi_s\right)_k$, $r = \overline{1, M}$, $s = \overline{1, M}$.
- Tables with the values of criteria and acceptable parameters and Pareto-optimal solutions, which allow the determination of how much of the obtained solution differs from what is required.

Step 6: Using visualization tools, re-solve a problem of optimization with regard to the modifications agreed with the participants of the product life-cycle who inputted data. The basis of the number of Pareto-optimal solutions found determines the most preferred solution.

The proposed algorithm is implemented in the form of applied computer programs "PVRK9". The program is created using the programming language MAPLE.

Let's consider the particular case when, in accordance with the requirements of the customer, the vessel must meet the following requirements and constraints. Allowable weight of not more than M_0 = 25 kg; volume V_0 = 0.04 m^3; working pressure not less than p = 30 MPa; the dimensions of the ballon must conform to the following parameters: R_0 = 0.15 m is outer radius; L_0 = 0.9 m is overall length. In the state of pin holes being fully closed ($k_{1,2}$ = 1), the shape of the bottom of the cylinder in the framework of the specified size is arbitrary. The initial data are $[\sigma_B]_{\pm 1}$ = 1600 and 600 MPa, $[\sigma_B]_{\pm 2}$ = 60 and 160 MPa is allowable stress in tension/compression of a composite material in the direction along/across the reinforcing fibres; E_1 = 48 GPa, E_2 = 10 GPa - modules of elasticity in the direction of the length and breadth of reinforcing elements of a layer, respectively, ν_{12} = 0.25 is Poisson ratio; δ = 0.005 m is the width of the tape; the ρ_{KM} = 2200 kg/m^3 density of the composite.

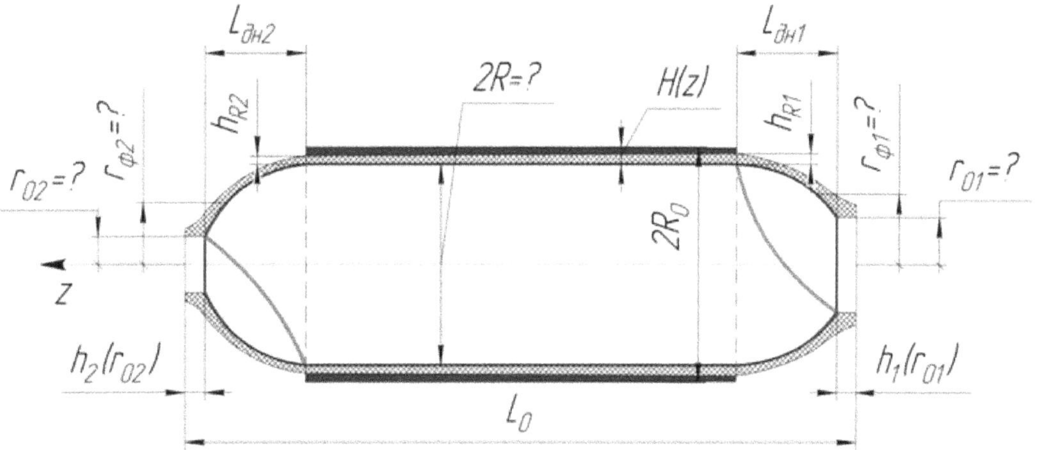

Figure 2. The geodesic winding scheme. The parameters to be determined are marked with a question mark. r_{01} – the radius of the pole hole of the first bottoms [m]; r_{02} – the radius of the pole hole of the second bottom [m]; $r_{\phi 1}$ – the radius of embedded element (flange) on the first bottom [m]; $r_{\phi 2}$ – the radius of embedded element (flange) on the second bottom [m]; R – the size of the internal profile of the composite pressure vessel [m]; $h_1(r_{01})$, $h_2(r_{02})$ –thickness of shell of vessel in vicinity of pole holes on the first and second bottom, respectively [m]; h_{R1}, h_{R2} – the thickness of the walls of the vessel during the first and second equator, respectively [m]; $L_{\partial н1}$, $L_{\partial н2}$ – the height of the first and second bottom (internal surface), respectively [m]; $H(z)$ – the projected thickness of the vessel on a cylindrical part as a function of z coordinate [m].

Table | Image | Options

	1	2	3	4	5	6	7	8
1	Ai	Phi1	Ai	Phi2	Ai	Phi3	Ai	Phi4
2	A7025	0.5448000000e-5	A8553	3.927146717506991e-5	A10200	19.433577708649686	A14537	0.0025591936186157354
3	A649	0.7398533333e-4	A8818	1.8368423325733352e-4	A16140	19.487112299397786	A16086	0.0034645807137398425
4	A7557	0.1094980000e-3	A6150	3.8744138745570866e-4	A8496	19.49283435004922	A11241	0.003932949068383557
5	A14460	0.1460280000e-3	A10041	5.832088112441708e-4	A6648	19.499338353592137	A11506	0.004142224519031975
6	A6014	0.1606933333e-3	A1702	6.403644552972071e-4	A14536	19.57831061112498	A3502	0.004155012747996496
7	A4614	0.3810046667e-3	A10030	8.886857376117635e-4	A3748	19.6011331737612	A3242	0.00493897863507192
671	A12218	0.5075081333e-1	A15973	0.13693903275364092	A7209	24.150505147727383	A8117	0.49431029310564845
672	A10200	0.5114240400e-1	A11649	0.13733730302607478	A15973	24.164440475527574	A9029	0.5008840314911679
673	A554	0.5125213400e-1	A3597	0.13833015899637305	A3597	24.237111690270137	A8481	0.5018457357519529
674	A16140	0.5176631267e-1	A10489	0.14162107045457786	A10489	24.276302435608677	A2617	0.5082032454034531
675	A7122	0.5215123267e-1	A15069	0.151175614120444	A15069	24.5384950947883	A10605	0.5364651455250193
676	A6648	0.5292671733e-1	A13389	0.15221887165167997	A13389	24.561108976623633	A14837	0.5564457843942214

Figure 3. Fragment of the table tests

Because of the availability of technological equipment for winding, the chosen technological method for manufacturing the pressure vessel selected was geodetic winding.

The calculation for the process of winding along a geodesic line is shown in Fig. 2 (the meaning of the values in the figure is explained below).

Here is the order of the solution in accordance with the methodology, which was set out in the paper.

1. Variable parameters and the constraints on them

Range 5 design parameters: r_{01}, r_{02}, $r_{\phi 1}$, $r_{\phi 2}$, R. Denote the vector of the parameters as α:

$$\alpha = (r_{01}, r_{02}, r_{\phi 1}, r_{\phi 2}, R) \equiv (\alpha_1, \alpha_2, \ldots, \alpha_5)$$

Parametric constraints are defined on the basis of construction and technological considerations. They are given in Table 1.

2. Functional dependencies and constraints on them

Enter the following functional dependence:

$$f_1(\overline{\alpha}) = r_{01} - r_{02}, f_2(\overline{\alpha}) = r_{\phi 1} - a_1 \cdot r_{01}$$

$$f_3(\overline{\alpha}) = r_{\phi 1} - \frac{R + 3 \cdot r_{01}}{4}, f_4(\overline{\alpha}) = r_{\phi 2} - a_2 \cdot r_{02}$$

$$f_5(\overline{\alpha}) = r_{\phi 2} - \frac{R + 3 \cdot r_{02}}{4}$$

$$a_i = \sqrt{\frac{3 + n(k_i - 4) + \sqrt{n^2(k_i^2 - 4k_i + 12) - 6n(n+2) + 1 + 8k_i}}{2(2-n)}}$$

$$n = \frac{E_2 + E_1 \cdot \upsilon_{12}}{E_1 \cdot (1 + \upsilon_{12})}; i = 1, 2.$$

Functional constraints are set by the following inequalities $f_1(\overline{\alpha}) > 0; f_2(\overline{\alpha}) \geq 0; f_3(\overline{\alpha}) \leq 0; f_4(\overline{\alpha}) \geq 0; f_5(\overline{\alpha}) \leq 0.$ They reflect the relationship between the structural dimensions of the projected container in accordance with the scheme shown in Fig. 2.

3. Performance criterions and a priori constraints

Let us enumerate the list of criteria for multi-criteria synthesis:

$\Phi_1(\overline{\alpha}) = \mathfrak{R}(\overline{\alpha})$ ≡ Rhoa(α) – the accuracy of the dimension

$\Phi_2(\overline{\alpha}) = \mathcal{V}(\overline{\alpha})$ ≡ Vhoa(α) – the accuracy of the volume

$\Phi_3(\overline{\alpha}) = \mathfrak{M}(\overline{\alpha})$ ≡ Mhoa(α) – weight

$\Phi_4(\overline{\alpha}) = |\tan \Psi_{II}|(\overline{\alpha})$ ≡ tanPsiTSC(α) – The feasibility of winding on the cylinder

The area of the target values of the criteria is as follows:

$$\overline{\Phi} = (0 \leq \Phi_1 = \text{Rhoa} \leq 10^{-3},\ 0 \leq \Phi_2 = \text{Vhoa} \leq 10^{-3},$$
$$0 < \Phi_3 = \text{Mhoa} \leq 25,\ 0 \leq \Phi_4 = \text{tanPsiTSC} \leq 0.35)$$

4. The original problem

The original range of parameter's variation (a parallelepiped P_0) is shown in Table 1:

Calculation of one vector of criteria is about 0.3 seconds of computer time. In a parallelepiped P_0 generated

Table 1

Variable parameters	Left boundary	Right boundary
r_{01}	0.03	0.07
r_{02}	0.03	0.07
$r_{\phi 1}$	0.03	0.07
$r_{\phi 2}$	0.03	0.07
R	0.129	0.1425

Ai	Φ1	Ai	Φ2	Ai	Φ3	Ai	Φ4
A4080	0.008827728000	A4080	0.00312519953851211	A4080	20.8864713975947	A4080	0.0790298384491721
A13192	0.005998748000	A13192	0.00408907407234774	A13192	21.0445116102266	A13192	0.0341438263782930
A15960	0.007284998000	A15960	0.00434073167175884	A15960	21.0418275388032	A15960	0.0963203146779909

Figure 4. The table criteria of the feasible solution set

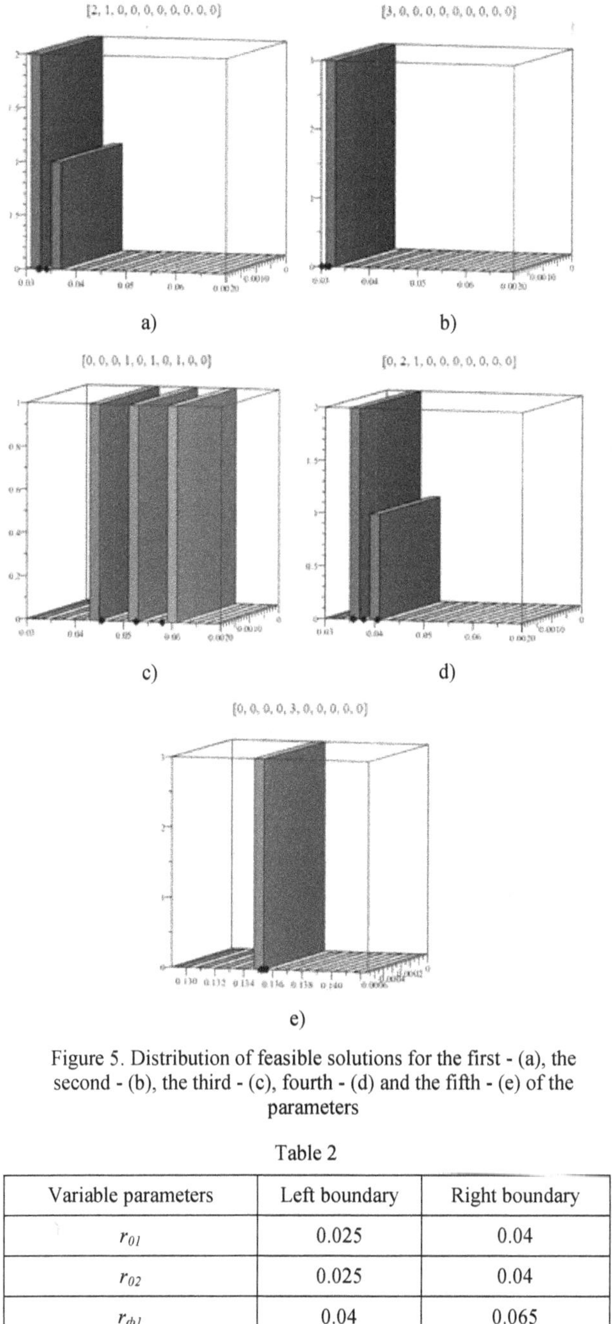

Figure 5. Distribution of feasible solutions for the first - (a), the second - (b), the third - (c), fourth - (d) and the fifth - (e) of the parameters

Table 2

Variable parameters	Left boundary	Right boundary
r_{01}	0.025	0.04
r_{02}	0.025	0.04
$r_{\varphi 1}$	0.04	0.065
$r_{\varphi 2}$	0.03	0.05
R	0.134	0.136

N_0 = 16384 sampling points by the LP$_\tau$ - sequence. After the calculation and verification of the functional constraints were M_0 = 675 vectors. On the basis of the received M_0 vectors, the calculated vector criteria were used to make the test table (Fig. 3). In the test table the values of criteria are organized in declining order (the best for each criterion vectors are in the upper part of the table).

After analyzing the table tests criterial constraints were set with a view to striving for data that achieved the optimal aims of; $[\Phi_1 \le 0.01; \Phi_2 \le 0.01; \Phi_3 \le 25; \Phi_4 \le 0.35]$. Three vectors #4080, #13192 #15960 were acceptable solutions. They also are Pareto-optimal (Fig. 4 shows the values of the criteria for these vectors).

From the table of criteria (Fig. 3) we see that criterion Φ_3 and Φ_4 already reach the desired value, but criterion Φ_1 and Φ_2 didn't yet.

Some of the distribution histograms of the above three solutions are shown in Fig. 5 (a, b, c, d, e). The variation interval of the parameter $[\alpha_j^*; \alpha_j^{**}]$ (j = 1..5) is divided into 10 equal parts. Over each interval appears the number indicating how many got feasible solutions there are. Black diamonds below shows the Pareto-optimal solutions.

The analysis of histograms and criteria table indicates the possible direction (black arrows in Fig. 5 a and b) further correction of parametric restrictions.

5. Correction of problem

Define new boundaries of a variation of parameters (parallelepiped P$_1$), the values of which are received in the result of the analysis of the initial problem (see Table 2).

Repeat the same procedure as the initial problem. In a parallelepiped P$_1$ generated N_1 = 2048 sampling points by the LP$_\tau$ - sequence. After the calculation and verification of the functional constraints are M_1 = 594 vectors. On the basis of the M_1 vectors expect vector criteria $\overline{\Phi}\left(\overline{\alpha}^{(i)}\right); i = \overline{1,594}$ and will make the table tests. And on the basis of tables of the test in this correction, were set more stringent criteria constraints $[\Phi_1 \le 0.005; \Phi_2 \le 0.0001; \Phi_3 \le 25; \Phi_4 \le 0.35]$. These restrictions have already coincided with the target values for the three criteria Φ_2, Φ_3 and Φ_4. It was found that only one vector #1830, satisfied all of the restrictions. Fig. 6 and 7 show the importance of criteria and parameters of this vector, respectively.

Ai	Φ1	Ai	Φ2	Ai	Φ3	Ai	Φ4
A1830	0.004695809333	A1830	0.0000616790269906964	A1830	20.9730486179006	A1830	0.0522106861431872

Figure 6. The value of the criteria of the vector 1830

Ai	α1	α2	α3	α4	α5
1830	0.03091064453	0.02950439453	0.05075439453	0.04875976562	0.1358134766

Figure 7. The value of the parameters of the vector 1830

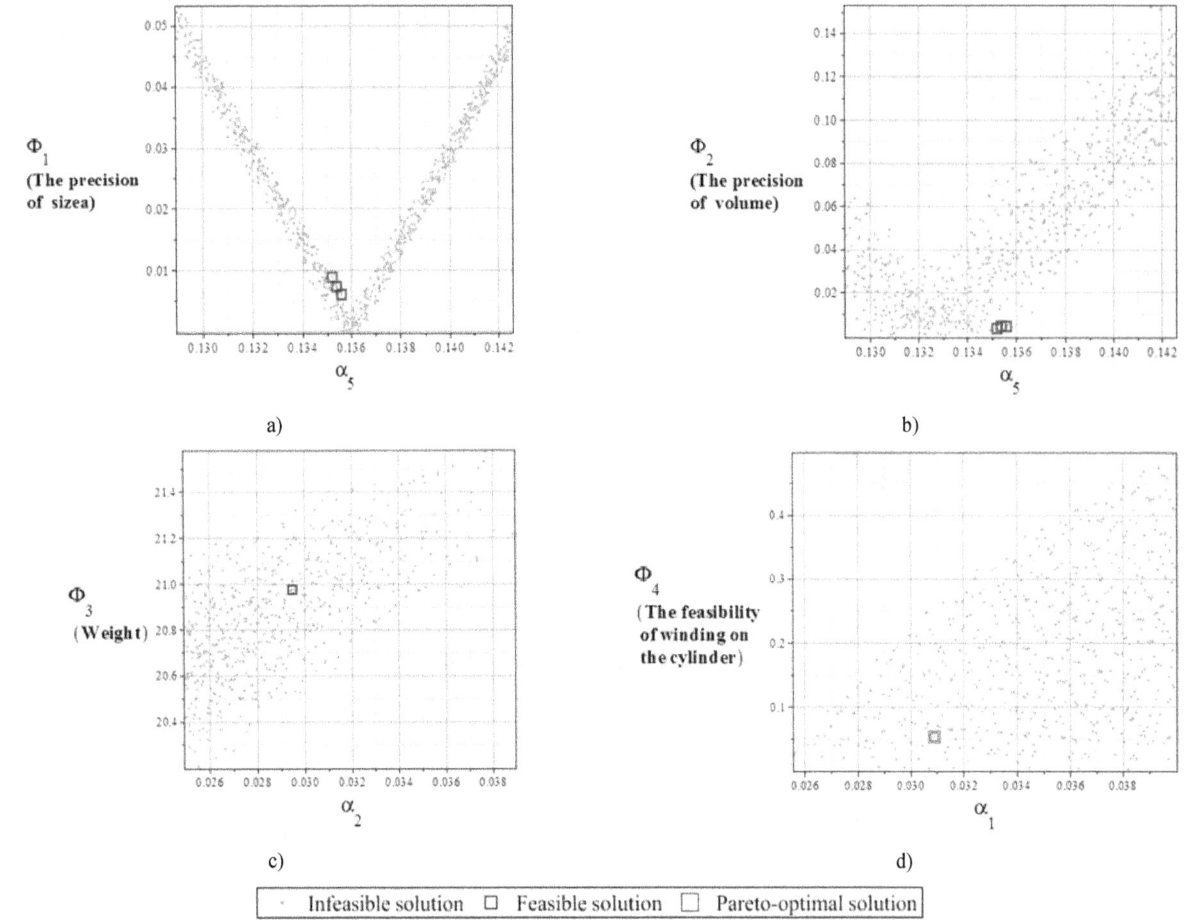

Figure 8. Plots of $\Phi_1(\alpha_5)$ (the original problem) – (a), $\Phi_2(\alpha_5)$ (the original problem) – (b), $\Phi_3(\alpha_2)$ (the correction) – (c), $\Phi_4(\alpha_1)$ (the correction) – (d)

6. Analysis of the results using visualization tools

Because it was not possible to reach the target values for all criteria within the context of the problem, it is necessary to use visualization tools to know how the criteria depend on the parameters, as well as the impact of each criterion on the other.

Fig. 8 (a, b, c, d) shows dependence criteria of some parameters. They are a projection of nine-dimensional points (five-dimensional parameter space and four-dimensional space-criteria) on the plane $\Phi_\nu\alpha_j$, $\nu = \overline{1,4}$, $j = \overline{1,5}$.

Fig. 9 (a, b, c, d) shows dependence criteria of some other criteria. They are a projection of nine- dimensional points (five-dimensional parameter space and four-dimensional space-criteria) on the plane $\Phi_i\Phi_j$, $i = \overline{1,4}$, $j = \overline{1,4}$.

Analysis of the data graph shows the complex dependencies of the criteria, parameters and between a criterion. However, due to the tolerances on the external dimensions of the resultant figure of 1830 we can get a solution that satisfies all the criteria at the same time. Namely, the value of $\Phi_1 = 0.004695809333$.This should mean that the size of the cylinder is equal to 0.14931. Replace the calculated size of 0.14931 into the mathematical model and solve the problem of finding admissible vectors in the region of parameters of vector 1830. Conducted 64 tests and the restrictions meet all the customer's requirements and are consistent with the capabilities of the performers. Found four admissible solutions. The preference was given to the vector 14 (see Fig. 10). The mass of the container reduced to 4 kg.

The values of the parameters of the vector 14 shown in Fig. 11.

This solved the problem of multi-criteria synthesis of the manufacturing process of a composite pressure vessel by the method of winding along a geodesic line. The mass of the container was reduced to 4 kg.

The method was developed for the search of acceptable and Pareto-optimal solutions for the synthesis of the process of winding pressure vessels from composites.

References

Bratukhin, A., Bogoliubov, V. and Sirotkin, O. (2003). *Technology of production and integral structures from composite materials in mechanical engineering*. Moscow: Gotika.

Bulanov, I., Smuislov, V. and Komkov, M. (1985). *Pressure vessels made of composite materials in the construction of aircraft*. Moscow: Central Scientific-Research Institute of Information.

Dang, M., Gavriushin, S. and Semisalov, V. (2012). Analysis and synthesis of the winding process of the composite vessel within the concept of product life-cycle management, *Proceedings of Higher Educational Institutions: Machine Building*, No 7, pp. 12-17.

Dang, M., Gavriushin, S. and Semisalov, V. (2012). Method for the synthesis process of winding composite cylinders within the concept product life-cycle management, *Science and Education*. [Online]. Available at http://technomag.edu.ru/doc/434726.html (accessed 28 February 2013).

Gavriushin, S. and Dang, M. (2012). Synthesis of composite pressure vessel within the concept of product life-cycle management, *Non-classical Problems of Mechanics*, pp. 326-330.

Komkov, M. and Tarasov, V. (2011). *Technology Winding Composite Constructions of Missiles and Weapons of Destruction*. Moscow: Bauman MSTU.

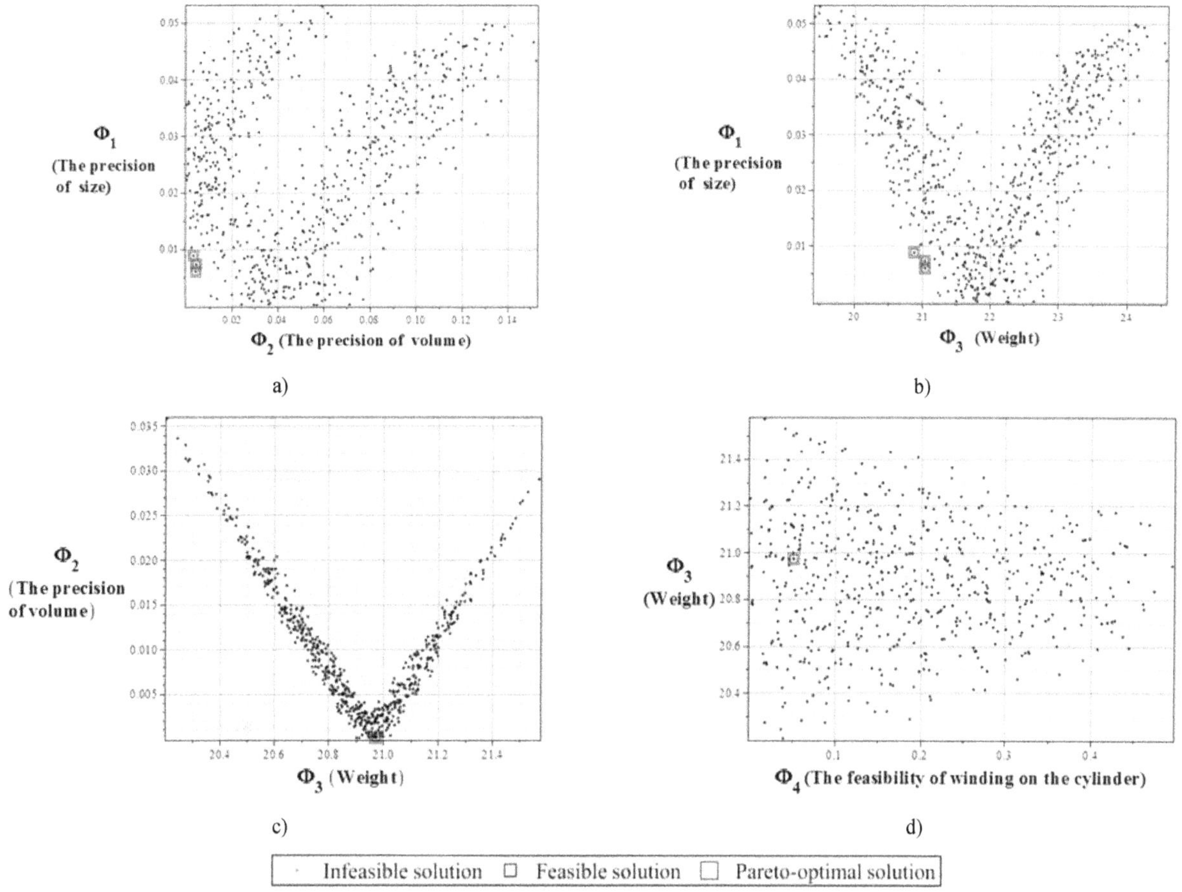

Figure 9. Plots of $\Phi_1(\Phi_2)$ (the original problem) – (a), $\Phi_1(\Phi_3)$ (the original problem) – (b), $\Phi_2(\Phi_3)$ (the correction) – (c), $\Phi_3(\Phi_4)$ (the correction) – (d)

Ai	Φ1	Ai	Φ2	Ai	Φ3	Ai	Φ4
A6	0.0003300020092	A6	0.000659042259594685	A6	20.9667773328739	A6	0.0615999183583240
A14	0.0001088714755	A14	0.000126093005362732	A14	20.9760165653525	A14	0.0438599856893639
A50	0.0008787469024	A50	0.000866684552289741	A50	20.9978621545466	A50	0.0393547488345502
A64	0.0009638972607	A64	0.000877707188905348	A64	20.9954895911663	A64	0.000885363708138731

Figure 10. Feasible solutions in the neighborhood of the vector 1830

Ai	α1	α2	α3	α4	α5
14	0.03087500000	0.02968750000	0.05012500000	0.04775000000	0.1358125000

Figure 11. The value of the parameters vector 14

Sarbaev, B. (2001). *The Calculation of the Power Envelope of the Composite Pressure Vessel*. Moscow: Bauman MSTU.

Sobol, I. and Statnikov, R. (2006). *The Selection of Optimal Parameters in Problems with Multi Criteria*. Moscow: Drofa.

Statnikov, R. and Matusov, I. (2012). About unacceptable, acceptable and optimal solutions in problems of design, *Problems of Mechanical Engineering and Reliability*, No 7, pp. 10-19.

Statnikov, R., Matusov, I. and Statnikov, A. (2012). Multicriteria engineering optimization problems: Statement, solution and applications, *Journal of Optimization Theory and Applications*. [Online]. Available at http://www.springerlink.com/content/1607x66142384577/, DOI: 10.1007/s10957-012-0083-9.

Statnikov, R. and Statnikov, A. (2011). *The Parameter Space Investigation Method Toolkit*, Boston, London: Artech House.

Vasiliev, V. (1988). *Mechanics of Structures From Composite Materials*. Moscow: Machine Building.

Analysis of Nonlinear Mechanical Behaviour Carbon Fibre Reinforced Polymer Laminates

Alexander Dumansky[1], Muchamyat Alimov[1], Ludmila Tairova[2]

[1] Institute of Machines Science of the Russian Academy of Sciences, 4 Maly Kharitonievsky per. Moscow, 101990, Russia
[2] Rocket and Spacecraft Composite Structures Department, Faculty of Special Machinery, Bauman Moscow State Technical University, 5 2nd Baumanskaya Street, Moscow, 105005, Russia

Abstract: An approach describing the nonlinear behaviour of multidirectional carbon fibre reinforced polymer composites has been elaborated as application of matrix algorithms to classical lamination theory relations. It is assumed that nonlinear in-plane shear strain arising in each lamina plays a dominant role in such nonlinear behaviour. Parameters of a model proposed have been defined through a two-stage procedure. The first stage corresponds to the field of elasticity and includes an estimation of the elastic engineering characteristics of the lamina by using an identification method. According to this method the lamina elastic characteristics have been defined on the basis of testing results for specimens with various lay ups. With this end in view the testing of specimens with 0, ±20°, ±40°, ±50°, ±70°, 90° lay ups under uniaxial tensile loading has been carried out. The second stage is associated with the field of nonlinear behaviour. Analytical approximation (including piecewise linear one) of the lamina shear stress-strain curve is used and substituted into the lamina stiffness matrix. It allows the stiffness matrix to be divided into two constituents describing consequently linear and nonlinear properties of the lamina. To obtain the stiffness matrix of the multidirectional laminate, the classical lamination theory relations have been used. Matrix algorithms have been applied for the purpose of performing the inverse of the stiffness matrix resulting in the explicit form of the compliance matrix. The engineering characteristics of the laminate may be obtained by using this approach. On the basis of the results received the anisotropy of nonlinear properties of carbon fibre reinforced composite laminates can be described. To illustrate the correctness of the approach, the stress-strain curves predicted for some lay ups have been compared with the experimental data and a satisfactory agreement has been observed. .

Key Words: Space antennas, Composite materials and structures, Metal meshes, Mathematical modelling, Tests.

1. Introduction

Predicting the mechanical behaviour of composite materials is very important because of their widespread use in aerospace, machine-building, automotive and other industries. Intensive experimental and theoretical investigations in the field of strength and failure of composite materials were conducted during the last decades. First of all, WWFE-World-Wide Failure Exercise (Hinton et al., 2004) should be noted. It was based on numerous papers of researchers from various countries. Most of the WWFE papers consider the unidirectional layer as a base for predicting the laminate properties, as well as the laminate non-elastic properties as a result of the nonlinear properties of the layer. As the nonlinear properties of the layer are induced by nonlinear properties under in-plane shear of the layer, it is important to get the shear stress-strain curve. There are different kinds of approximation of such curves (Soden et al., 2004; Kaddour et al., 2011; Hinton et al., 1996; Bogetti et al., 2004; Hinton et al., 1969; Hahn and Tsai, 1973; Rosen, 1972; Zinoviev et al., 2004a; Zinoviev et al., 2004b), but the principal idea of the approach proposed in this paper is to separate of elastic strain from the non-elastic one.

As to the elasticity range, it is important to estimate correctly the lamina mechanical properties. This problem may be successfully solved with the aid of the identification method (Dumansky et al., 2011). The method is based on minimization of residual between experimental and calculated data. Testing of various angle-ply laminates is preferable to define experimentally the lamina properties (Hinton et al., 1969; Bogetti et al., 2004; Dumansky et al., 2011; Alimov et al., 2011). It should be noted that significant nonlinearity of stress-strain curves is observed in case of loading not coinciding considerably with the fibre direction of any of the layers (Hinton et al., 1996; Bogetti et al., 2004; Hinton et al., 1969; Hahn and Tsai, 1973; Rosen, 1972; Zinoviev et al., 2004a; Zinoviev et al., 2004b; Dumansky et al., 2011; Alimov et al., 2011). In some cases (Bogetti et al., 2004; Zinoviev et al., 2004a) such nonlinearity is taken into account by applying the incremental algorithms to stresses or strains. It is possible to use analytical approximation of the stress-strain curves as in (Bogetti et al., 2004) where the Ramberg-Osgood nonlinear three-parameter equation was applied. There are some other attempts to describe the lamina nonlinear behaviour (Hinton et al., 1969; Rosen, 1972; Zinoviev et al., 2004a), but elaborating the methods suitable for practical use and providing satisfactory agreement with experimental data is still of current importance.

2. Identification of lamina elastic properties

Using the identification method is preferable for defining the lamina properties of angle-ply specimens in comparison with the method based on testing of unidirectional laminates under off-axis loading. In the general case the problem of identification is reduced to minimization of a residual function providing for, to some extent, the minimization between theoretical and experimental data. In particular, the determination of the lamina properties can be carried out by the use of testing data of angle-ply lay ups, strains being measured in both longitudinal and transverse directions. This can be expressed by the following relation (Dumansky et al., 2011)

$$\min_{g_{ij}^0}\left\lfloor \sum\left(\varepsilon_x^{calc}-\varepsilon_x^{\exp}\right)^2+\left(\varepsilon_y^{calc}-\varepsilon_y^{\exp}\right)^2 \right\rfloor, \qquad (1)$$

where g_{ij}^0 are the lamina stiffness components in the main axes of orthotropy; e_x^{calc}, e_y^{calc}, e_x^{exp}, e_y^{exp} are the calculated and experimental strains of angle-ply laminate correspondingly.

In case of uniaxial loading the matrix form of constitutive equations is

$$\begin{Bmatrix} \varepsilon_x \\ \varepsilon_y \\ 0 \end{Bmatrix} = \begin{pmatrix} s_{xx} & s_{xy} & 0 \\ s_{xy} & s_{xx} & 0 \\ 0 & 0 & s_{ss} \end{pmatrix} \begin{Bmatrix} \sigma_x \\ 0 \\ 0 \end{Bmatrix} \qquad (2)$$

where s_{xx}, s_{xy}, s_{yy}, s_{ss} are the compliance components of angle-ply laminate.

Since the matrices of compliance and stiffness are connected by the equality $[S_{xy}] = [G_{xy}]^{-1}$ the relations for the strain may be written in the following form:

$$\varepsilon_x^{calc} = \frac{g_{xx}}{D_{xy}} \sigma_x; \qquad \varepsilon_y^{calc} = -\frac{g_{xy}}{D_{xy}} \sigma_x \qquad (3)$$

where $D_{xy} = g_{xx} g_y - g_{xy}^2$.

In its turn the relations of stiffness components of angle -ply laminate $[\pm\theta]$ may be written as (Hinton et al., 1996).

$$g_{xx} = g_{11}^0 \cos^4\theta + g_{22}^0 \sin^4\theta + 2\left(g_{12}^0 + 2g_{66}^0\right)\sin^2\theta\cos^2\theta$$

$$g_{yy} = g_{11}^0 \sin^4\theta + g_{22}^0 \cos^4\theta + 2\left(g_{12}^0 + 2g_{66}^0\right)\sin^2\theta\cos^2\theta$$

$$g_{yy} = \left(g_{11}^0 + g_{22}^0 - 4g_{66}^0\right)\sin^2\theta\cos^2\theta + \left(\sin^4\theta + \cos^4\theta\right)g_{12}^0 \qquad (4)$$

where $g_{11}^0 = \dfrac{E_1}{1-\nu_{12}\nu_{21}}$, $g_{22}^0 = \dfrac{E_2}{1-\nu_{12}\nu_{21}}$, $g_{12}^0 = \dfrac{\nu_{21}E_1}{1-\nu_{12}\nu_{21}}$, $g_{66}^0 = G_{12}$ and E_1, E_2, G_{12}, ν_{12} are the lamina engineering constants.

Thus equations (1)-(4) allow all the lamina elastic properties to be defined.

3. Non-linear shear stress-strain curve approximation

The experimental results allow us to assume that for the lamina elastic moduli in the main orthotropy axes E_1, E_2 and Poisson's ratio ν_{12} may be taken invariant. The reason of nonlinear behaviour is the lamina in-plane shear properties. The shear stress-strain curves may be given in advance as in the WWFE instruction or defined from the indirect experimental data. These curves may be approximated by a function (for example by piecewise linear one). A diagram in the piecewise form is shown in Fig. 1.

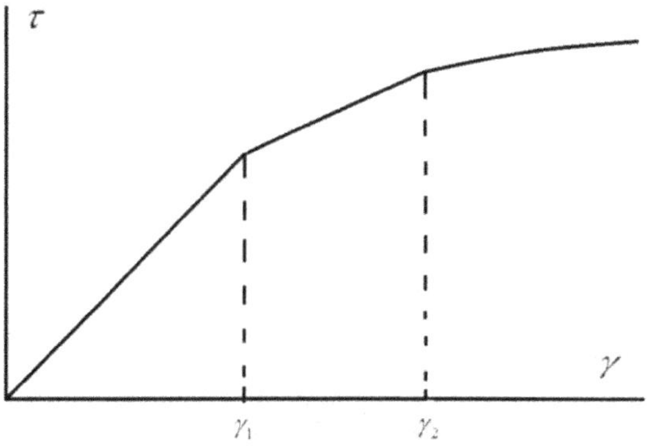

Figure 1. Lamina in-plane shear stress-strain approximation by piecewise function.

Such an approximation may be analytically expressed as follows

$$g_{66}(\gamma) = g_{66}^0 - \Delta g_{66}^{(1)} H(\gamma - \gamma_1) - \Delta g_{66}^{(2)} H(\gamma - \gamma_2) \ldots \qquad (5)$$

or in case of approximation of the nonlinear part of the stress-strain curve by power function $a(\gamma - \gamma_1)^n$

$$g_{66}(\gamma) = g_{66}^0 - an(\gamma - \gamma_1)^{n-1} H(\gamma - \gamma_1) \qquad (6)$$

where g_{66}^0 is shear modulus of the linear area, $\Delta g_{66}^{(k)}$ characterizes change in shear modulus of the r-*th* part of the strain axis, $H(\ )$ is Heaviside unit function, a, n are parameters of the power function. It should be noted for equation that the material reaches the yield area when $\sum_{k=1}^{n} \Delta g_{66}^{(k)} = g_{66}^0$

Experimental shear stress-strain curves for the materials given to contributors of the third world-wide failure exercise (WWFE-III) Part (A) (Soden at al., 2004) are shown in Fig. 2.

As is seen from Fig.2 the lamina shear-stress strain curves are essentially nonlinear and it is possible to represent them by equations (5) or (6). Two or three linear parts may be sufficient to describe the lamina shear stress-strain curve (Dumansky and Tairova, 2008). In the general case, the approximation by equations (5) and (6) may be rewritten in the following form

$$g_{66}(\gamma) = g_{66}^0 - f(\gamma) \qquad (7)$$

Then the lamina stiffness matrix may be expressed as

$$\left[G_{12}\right] = \begin{pmatrix} g_{11}^0 & g_{12}^0 & 0 \\ g_{12}^0 & g_{22}^0 & 0 \\ 0 & 0 & g_{66}^0 \end{pmatrix} - \begin{pmatrix} 0 & 0 & 0 \\ 0 & 0 & 0 \\ 0 & 0 & 1 \end{pmatrix} f \qquad (8)$$

For the purpose of matrix calculus equation may be represented as

$$\left[G_{12}\right] = \left(G_{12}^0\right) - \left[I_0\right] f \qquad (9)$$

Matrix stiffness of the laminate is formed in accordance with classical lamination theory relations (Rabotnov, 1979)

$$\left[G_{xy}\right] = \sum_k \left[T_1^{(k)}\right]\left[G_{12}\right]\left[T_1^{(k)}\right]^T \bar{h}^{(k)} \qquad (10)$$

where

$$\left[T_1^{(k)}\right] = \begin{pmatrix} c_{(k)}^2 & s_{(k)}^2 & -2s_{(k)}c_{(k)} \\ s_{(k)}^2 & c_{(k)}^2 & 2s_{(k)}c_{(k)} \\ s_{(k)}c_{(k)} & -s_{(k)}c_{(k)} & c_{(k)}^2 - c_{(k)}^2 \end{pmatrix} \qquad (11)$$

is transformation matrix, $s_{(k)} = \sin\theta_k$, $s_{(k)} = \cos\theta_k$, θ_k is angle between Ox axis and the fibre direction in the k-th lamina.

Substituting the stiffness matrix of equation (9) into (10) yields

$$\left[G_{xy}\right] = \left[G_{xy}^0\right] - \left[\tilde{(}\ \ \right] \qquad (12)$$

where $\left[G_{xy}^0\right] = \sum_k \left[T_1^{(k)}\right]\left[G_{12}^0\right]\left[T_1^{(k)}\right]^T \bar{h}^{(k)}$,

$\left[\tilde{(}\ \ \right]\ \ \_\ T_1^{(k)}\big]\left[I_0\right]\left[T_1^{(k)}\right]^T \bar{h}^{(k)}$.

Thus, the explicit form of the stiffness matrix has been obtained.

It is necessary also to obtain the compliance matrix. The matrix of compliance is the stiffness matrix inverse.

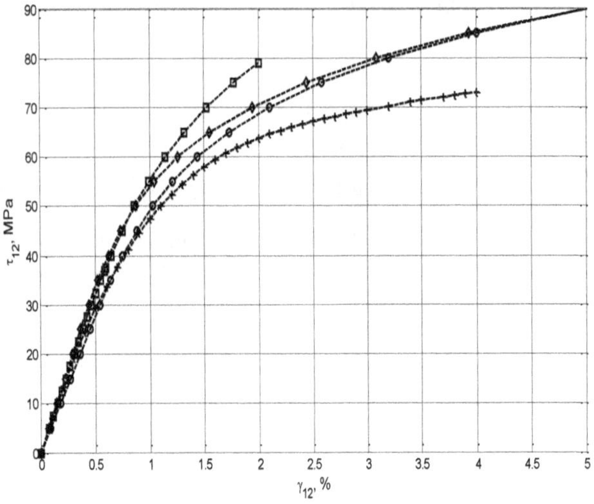

Figure 2. Longitudinal lamina shear stress-strain curves. Glass/LY556 (crosses), carbon/IM7/8552 (circles), G40-800/5260 (diamonds), AS4/3501-6 (squares).

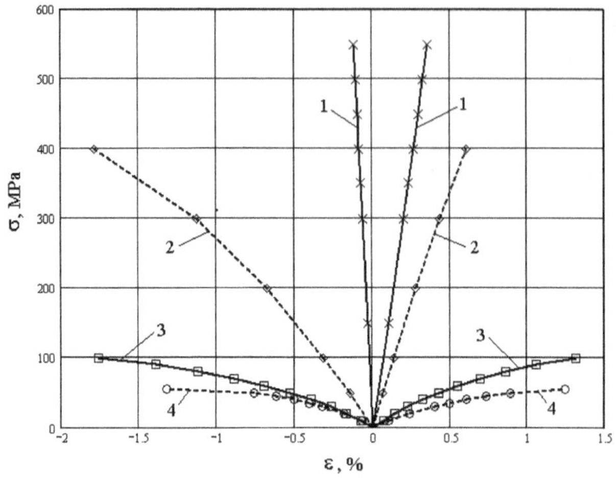

Figure 3. Longitudinal and transverse stress-strain curves for 1 - $[0]_4$; 2 - $[\pm20]_2$; 3 - $[\pm40]_2$; 4 - $[\pm50]_2$ lay-ups.

The compliance matrix may be presented as follows (Dumansky et al., 2011)

$$\left[S_{xy}\right]=\left[R\right]diag\left(\frac{1}{1-\lambda_i f}\right)\left[R\right]^{-1}\left[S_{xy}^{0}\right], \tag{13}$$

where λ_i are the eigenvalues of matrix $\left[G_{xy}^{0}\right]^{-1}\left[\tilde{G}\right.$ matrix $[R]$ is obtained as a result of matrix diagonalization: $\left[G_{xy}^{0}\right]^{-1}\left[\tilde{G}\right. \ldots diag\left(\lambda_1 \;\; \lambda_2 \;\; \lambda_3\right)\left[R\right]^{-1}$. With the aid of (13) the engineering constants of the laminate may be obtained (Dumansky et al., 2011) and the compliance matrix takes the following form:

$$\left[S_{xy}\right]=\begin{pmatrix} \frac{1}{E_x} & -\frac{\nu_{yx}}{E_y} & 0 \\ -\frac{\nu_{xy}}{E_x} & \frac{1}{E_y} & 0 \\ 0 & 0 & \frac{1}{G_{xy}} \end{pmatrix} \tag{14}$$

The eigenvalues are defined by lay-up of the laminate; function f characterizes the in-plane shear properties degradation.

4. Experimental procedure

To illustrate the approach, flat specimens with $[0]_4$, $[\pm20]_2$, $[\pm40]_2$, $[\pm50]_2$, $[\pm70]_2$, $[90]_4$, lay-ups were tested under quasi-static tensile loading. Minimizing equation (1) within the elastic strain area the lamina engineering constants were defined. The most of strain nonlinearity is observed for $[\pm40]_2$, $[\pm50]_2$ lay-ups which is induced by significant in-plane shear strain of the lamina. Longitudinal and transverse stress-strain curves are presented in Fig. 3, 4.

5. Model of Application

Elastic constants defined by means of the identification method are as follows: E_1 = 184 GPa, E_2 = 11.1 GPa, E_2 = 6.5 GPa, ν_{12} = 0.33. Nonlinear in-plane shear of the lamina is defined by the following equation:

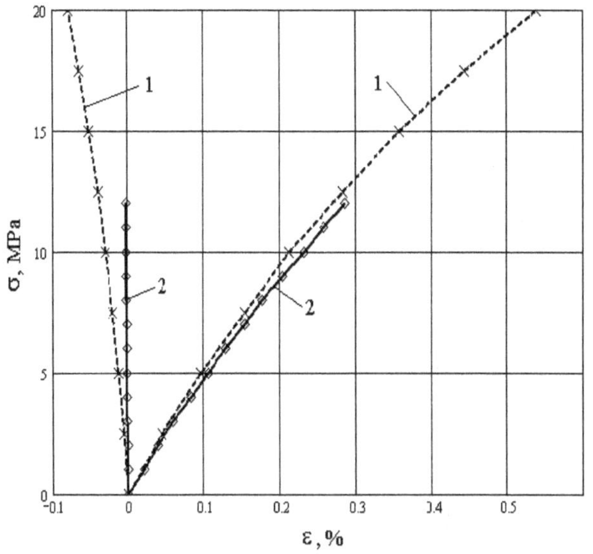

Figure 4. Longitudinal and transverse stress-strain curves for 1 - $[\pm70]_2$, 2 - $[90]_4$ lay-ups.

$$\tau(\gamma)=g_{66}^{0}\gamma-\left\lfloor g_{66}^{0}\left(\gamma-\gamma_*\right)-\left(\gamma-\gamma_*\right)^{n}\right\rfloor H\left(\gamma-\gamma_*\right) \tag{15}$$

where γ_* = 0.7% is the threshold value of shear strain corresponding to the beginning of nonlinearity, g, γ are the parameters of the power function approximating the nonlinear strain. They are as follows: g = 2.2 GPa, n = 0.95. Then the shear modulus approximation may be written as

$$g_{66}=g_{66}^{0}-\left\lfloor g_{66}^{0}-g\left(\gamma-\gamma_*\right)^{n-1}\right\rfloor H\left(\gamma-\gamma_*\right) \tag{16}$$

In case of linear approximation the shear modulus is stated as

$$g_{66}=g_{66}^{0}-\Delta g H\left(\gamma-\gamma_*\right) \tag{17}$$

In equation (17) $\Delta g \approx 5.4$ GPa.

Using the compliance matrix of expression (13) the longitudinal and transverse strains may be calculated as

$$\begin{aligned}\varepsilon_x^{(i+1)}&=\varepsilon_x^{(i)}+s_{xx}^{(i)}\left(\sigma_x^{(i+1)}-\sigma_x^{(i)}\right)\\ \varepsilon_y^{(i+1)}&=\varepsilon_y^{(i)}+s_{xy}^{(i)}\left(\sigma_x^{(i+1)}-\sigma_x^{(i)}\right)\end{aligned} \tag{18}$$

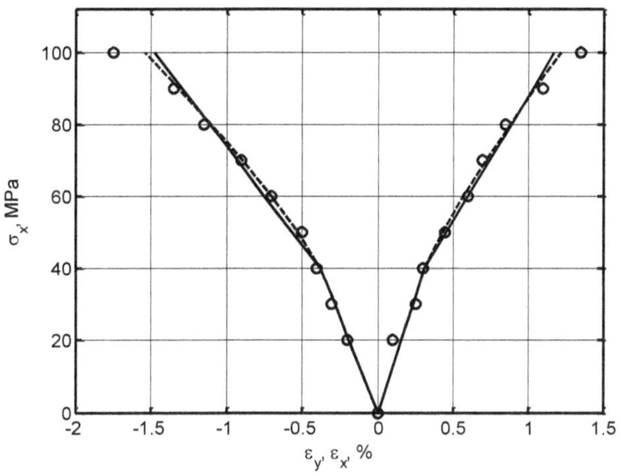

Figure 5. Comparison of the experimental data (circles) and calculated stress-strain curves for $[\pm 40]_2$ lay-up. Solid line is calculated by using piecewise linear approximation of shear stress-strain curve, dashed line for power one.

Comparison of experimental and calculated data is shown in Fig. 5.

6. Conclusion and recommendations

The analytical model proposed on the basis of matrix algorithms and classical lamination theory satisfactorily describes the nonlinear stress-strain response of carbon fibre reinforced laminates.

Applying the matrix algorithms to the classical lamination theory relations has proved to be highly effective in obtaining the matrices of stiffness and compliance of the laminates that allow the nonlinear mechanical properties to be described.

The approach may be generalized and extended to prediction of the rheological properties of carbon fibre reinforced plastics, as well as applied for describing the behaviour of laminates under the time variable loadings. On the basis of algebra of resolvent operators [(Rabotnov, 1979) a matrix method for the construction of hereditary constitutive equations was elaborated (Dumansky and Tairova, 2008). Combining such a matrix method with the approach proposed in this paper allows the anisotropy of nonlinear and rheological properties to be described.

The approach proposed may be used in strength analysis and design of thin-walled structures of composite laminates.

References

Hinton, M.J., Kaddour, A.S. and Soden, P.D. (Eds) (2004). *Failure Criteria in Fibre Reinforced Polymer Composites. The World-Wide Failure Exercise*, Elsevier.

Soden, P.D., Kaddour, A.S. and Hinton, M.J. (2004). Recommendations for designers and researchers resulting from the world-wide failure exercise, *Composites Science and Technology*, Vol. 64, pp. 589-604.

Kaddour, A.S., Hinton, M.J. and Soden, P.D. (2011). Damage prediction in polymeric composites: Up-date of part (A) of the third world-wide failure exercise (WWFE III), *Proceed. 18th International Conference on Composite Materials*, QinetiQ Ltd.

Hinton, A.S., Kaddour, P.D. and Soden P.D. (1996). Strength of composite laminates under biaxial loads, *Applied Composite Materials*, Vol. 3, pp. 179-198.

Bogetti, N.A., Hoppel, C.P.R., Harik, V.M., Newill, J.F. and Burns, B.P. (2004). Predicting the nonlinear response and progressive failure of composite laminates, *The World-Wide Failure Exercise*. Hinton, M.J., Kaddour, A.S. and Soden, P.D. (Eds), Elsevier, pp. 402-428.

Petit, P.H. and Waddoups. M.E. (1969). A method of predicting the nonlinear behaviour of laminated composites, *Journal of Composite Materials*, Vol. 3, pp. 2-19.

Hahn, H.T. and Tsai, S.W. (1973). Nonlinear elastic behaviour of unidirectional composite laminae, *Journal of Composite Materials*, Vol. 7, pp. 102-118.

Rosen, B.W. (1972). A simple procedure for experimental determination of the longitudinal shear modulus of unidirectional composites, *Journal of Composite Materials*, Vol. 6, pp. 552-554.

Bogetti, T.A., Hopplel, C.P.R., Harik, V.V., Newill, J.F. and Burns, B.P. (2004). Predicting the nonlinear response and progressive failure of composite laminates, *Failure criteria in fibre reinforced polymer composites. The World-Wide Failure Exercise*, Hinton, M.J., Kaddour, A.S. and Soden, P.D. (Eds), Elsevier, pp. 402-428.

Zinoviev, P.A., Grigoriev, S.V., Lebedeva, O.V. and Tairova, L.P. (2004a). The strength of multilayered composites under a plane stress -state, *Failure criteria in fibre reinforced polymer composites. The World-Wide Failure Exercise*, Hinton, M.J., Kaddour, A.S. and Soden, P.D. (Eds), Elsevier, pp. 379-401.

Zinoviev, P.A., Lebedeva, O.V. and Tairova, L.P. (2004b). A coupled analysis of experimental and the theoretical results on the deformation and failure of composite laminates under a state of plane stress, *Failure criteria in fibre reinforced polymer composites. The World-Wide Failure Exercise*, Hinton, M.J., Kaddour, A.S. and Soden, P.D. (Eds), Elsevier, pp. 943-960.

Dumansky, A.M., Tairova, L.P., Gorlach, I. and Alimov, M.A. (2011). Analytical and experimental study of nonlinear properties of carbon plastics, *Journal of Machinery Manufacture and Reliability*, Vol. 40, No 5, pp 483-488.

Alimov, M.A., Dumansky, A.M. and Radchenko, A.A. (2012). Analysis of nonlinear behaviour of angle-ply plastics under uniaxial tensile loads, *Journal of Machinery Manufacture and Reliability*, Nol. 41, No 2, pp. 132-136.

Rabotnov, Yu.N. (1979). *Elements of Hereditary Solids Mechanics*. Mir.

Dumansky, A.M. and Tairova, L.P. (2008). A method for the construction of hereditary constitutive equations of laminates bases on a hereditary constitutive equation of a layer, *Proceed. AIP Conference Current Themes in Engineering Science*, Korsunsky, A.M. (Ed.), Vol. 1045, Melville, New York, pp.71-80.

Photoluminescence Spectroscopy of Nematic Liquid Crystal Molecules

Shaimaa M. Abd Al-Baqi, Nasreen R. Jber, Ziad T. Al-Dahan, Fahad M. Abd Al-Hussain

Alnahrain University, Baghdad, Iraq

Abstract: This work based on testing the absorption spectra of a nematic mesophase of new synthesized material (4-ethoxybenyzililidene-4'-phenylazobenzene) soluble in Ethanol by using far infrared transmission measurement (FTIR) and analyze the peaks in these spectra. Fluorescence spectra were recorded from these molecules by using excitation wavelengths corresponding to the peaks of their absorption diagrams by using a fluorescence spectrometer and spectrofluorometer at the ultraviolet and visible range. As the molecule emit at a wide band in the visible range, it is a promising material at the field of optoelectronic material.

Key Words: Photoluminescence, 4-ethoxybenyzililidene-4'-phenylazobenzene, FTIR.

1. Introduction

Semiconductor devices have vastly influenced technological progress. This technology includes wide range of applications like; information exchange, TV displays and medical applications which have already been published by Baldo et al. (1998), Dimitrakopoulos and Malenfant (2002), and Muccini (2006). Alternative materials like organic semiconductors have received special attention in recent years because of their low production cost and easy processing. These organic materials are the basis of organic electronics, optoelectronics, resistors, capacitors, light-emitting diodes, field-effect transistors and optically pumped solid-state lasers can be fabricated by different methods and integrated into electronic and optoelectronic circuits (Huang and Chen, 2002) and (Peumans et al., 2003). It is necessary to realize the optical properties for the new synthesized materials which can be examined by using far infrared and photoluminescence measurements. All the atoms in molecules are in continuous vibration with respect to each other at temperatures above absolute zero. When the frequency of the IR radiation is equal to a specific vibration frequency at the molecules, the molecule absorbs the radiation. The excited molecule then decays back to the ground state, or to a lower-lying excited vibrational state, by emission of light. The emitted light is detected. Infra-red spectrum gives information about functional groups (Tiong et al., 2011), energetic levels and the reaction tendency. While the absorption and fluorescence spectra decide an optical frequency range of operation for materials and their applications at the optoelectronic technology. Photoluminescence is a type of optical spectroscopy in which ultraviolet, visible, or near infrared radiation is absorbed by a molecule to give information about the electronic and vibrational frequencies (vibronic transitions) (Settle, 1997). However, most types of CM continue to be more expensive than the traditional metals and alloys, composite technology is characterized by low energy efficiency, and the equipment is extremely complex and material intensive. The development of new technologies and there transfer from laboratory to manufacture is still largely done by means of intuitive-empirical methods.

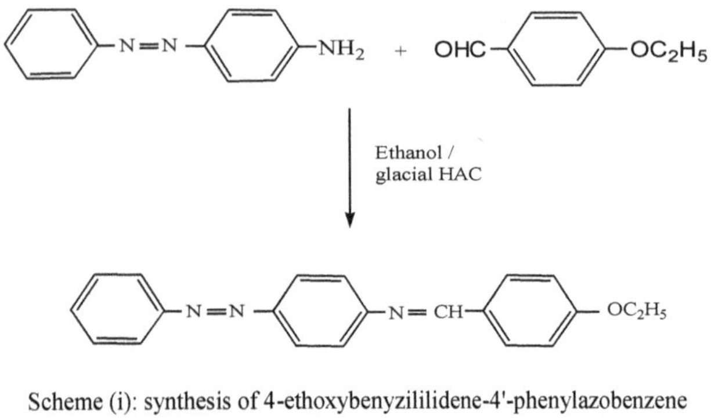

Scheme (i): synthesis of 4-ethoxybenyzililidene-4'-phenylazobenzene

2. Experimental work

All chemicals and solvents were of analar grade and were used without further purification. The identities of the prepared compounds were confirmed by the measurement of their infrared spectra (8300 shimadzo spectrophotometer). The transition temperatures were recorded by hot stage polarized microscope (Meiji MT9000). 4-ethoxybenyzililidene-4'-phenylazobenzene was synthesized according to scheme (i).

Schiffs bases were prepared by refluxing amino azobenzene (I) (0.01 mol, 1.97 g) with appropriate aldehydes (0.01 mole) with two drops of glacial acetic acid were heated under reflux in ethanol for 3-4 hrs as it shows by Suprvnowicz et al. (1985). The Schiff's bases crystallized out on cooling and were recrystallised from ethanol for purification. Give melting point (120-122°C) with yield 90%.

Absorption and Photoluminescence spectrum were studied for the synthesized material to determine the interesting spectral range of operation of 4-ethoxy benzilidine-4'-phenyl to select them for the specific optoelectronic devices.

At the beginning, it was important to determine the absorption spectrum for the selected material using spectrophotometer. Then, the fluorescence spectra were measured using fluorescence spectrometer (F96-pro spectrofluorometer). The spectra were recorded by measuring the emissions from the molecule which have been excited either by the whole range radiation directly from the lamp or using different filters. Finally, it was necessary to analyze these spectra to decide the important range for further examinations and future applications.

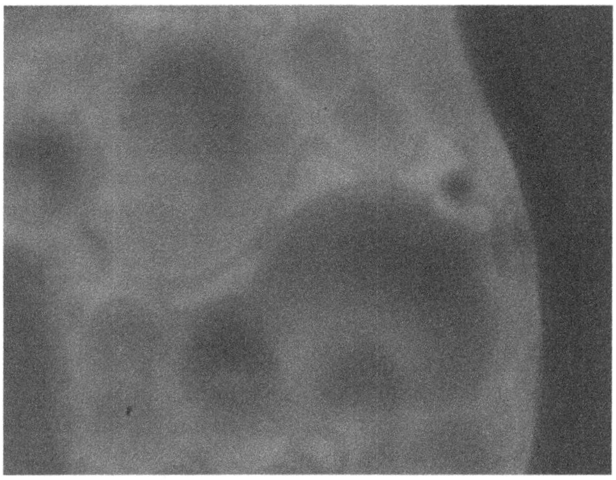

Figure 1. Nematic texture of 4-ethoxy benzilidine-4'-phenyl azobenzene as viewed under optical microscope at 200X

3. Results and disscusuion

It is well known that the type of mesophase (smectic or nematic) is determined mainly by the intermolecular attractions which operate between the slides and planes of the molecules, i.e., the strength of the lateral and terminal attraction forces (Lambert et al., 2011).The synthesized compound shows Schlieren and droplet-like nematic texture, as show in Fig. 1.

The compound was identified by FTIR spectroscopy. The spectrum in Fig. 2 (KBr disc cm^{-1}) shows the disappearance of the NH_2 group and the appearance of new band at 1605.1 cm^{-1} which was assigned to the (HC = N) group. The spectrum also show bands at 2939.2 cm^{-1} and 2857.6 cm^{-1} which corresponds to asymmetrical and symmetrical frequencies, while frequencies detected at 1512.6 cm^{-1}, 1020.4 cm^{-1}, 1597.8 cm^{-1} and 846.1 cm^{-1} are related to aliphatic (C – H) stretching, N = N stretching, C – C, C = C stretching and out of plane bending of *p*- substituted respectively (Sharma, 2012).

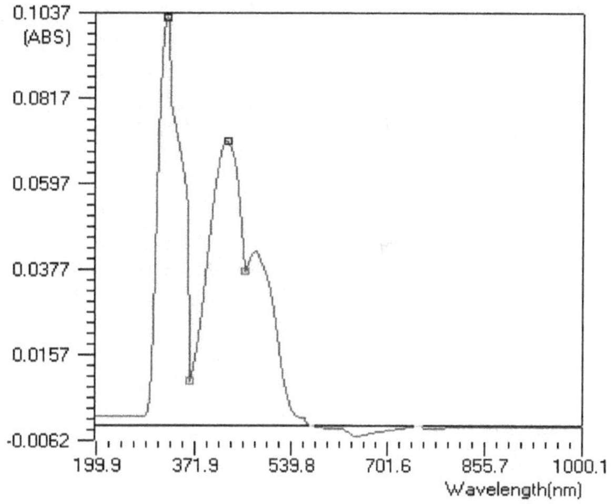

Figure 3. The absorption spectrum for the first sample (4-ethoxybenyililidene-4′-phenylazobenzene)

The absorption spectra for (4-ethoxybenyililidene-4′-phenylazobenzene) were recorded by using photospectrometer in the range (190-1200 nm) at 300K as shown in Fig. 3. It can be seen that the absorption spectra of the studied sample is mainly in the ultraviolet range and short wavelengths from the visible range. The absorption spectrum was measured in comparison to a reference sample (Ethanol). The absorption spectrum has three species. These species correspond to the vibronic peaks.

The reason of the broadening spectra in the studied sample initiate from the fact the molecule has high molecule weight because it consists of large numbers of carbon and hydrogen atom. The interaction between the ethanol and the molecules could also cause a broadening spectrum.

The photoluminescence spectra or the fluorescence

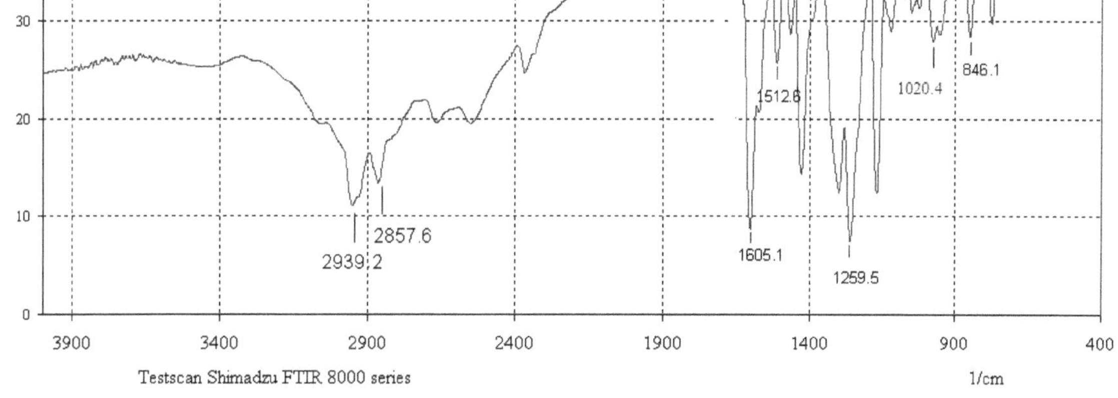

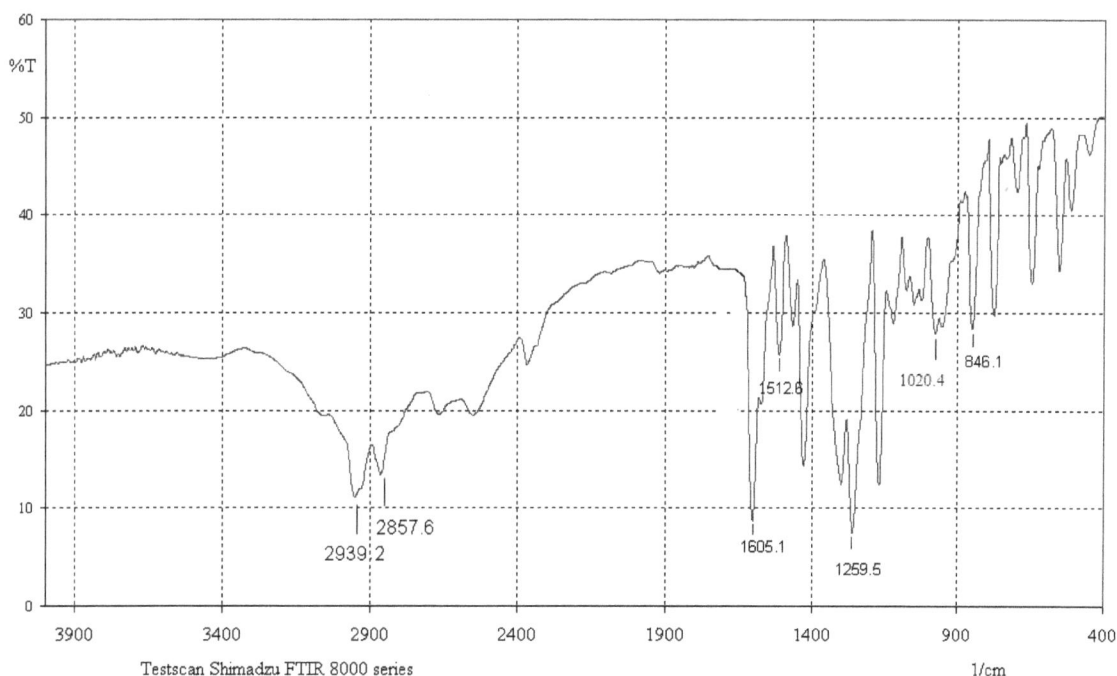

Figure 2. FTIR spectrum of 4-ethoxy benzilidine-4'-phenyl azobenzene

measurements have been studied using the fluorescence spectrometer in the range (200-900 nm) at 300K. Fig. 4 shows the emission spectrum for the Xe lamp used to excite the studied samples. If we compared this spectrum by the spectra emitted from the sample as shown in Fig. 5, it is observed that the molecule highly absorb at the UV and blue range. There is high emission spectrum at the visible range between (480-800 nm). While the absorption edge end at 570 nm. There is overlap between the absorption spectrum and the emission spectrum is (15.1 e.V) corresponding to the decay within the higher energy band of molecules. The overlap and the decay are evidences that 4-ethoxybenyililidene-4′-phenylazobenzene is a semiconductor material which has optical property similar to other semiconductor molecules are shown at the publications of Petrenko et al. (2009) and Da Silva Filho et al. (2005).

The black circle in Fig. 5 assigns a peak at the infrared which is corresponding to the same emission from the source. This emission is not related to the emission spectrum of the sample.

The reason behind this overlap can be either the effect of the interaction between the molecule and the solvent (Ethanol) or because a strong self-absorption which is corresponding to a broad band (Gommans, 2009).

Therefore, it was necessary to use filters in different range to reduce the confusing concerning the overlap between the absorption and the emission spectra. Fig. 6 and Fig. 7 show the spectra emission from the filter and the sample. The filters show two emission peaks; the main range has $\lambda_p = 365$ nm which correspond to the absorption range of the material, while the other peak $\lambda_p = 725$ nm. The sample shows a weak emission at the range (490-680 nm) which confirms the emission spectrum of the sample.

Conclusion

The emission and absorption spectra are similar to the spectra obtained from the organic molecules with semiconductor properties. The functional material 4-ethoxybenyililidene-4′-phenylazobenzene shows clear absorption and emission spectra with small overlap range. The overlap corresponds to the tail at the density of state of the highest unoccupied level.

References

Baldo, M.A., O'Brien, D.F., You, Y., Shoustikov, A., Sibley, S., Thompson, M.E. and Forrest, S.R. (1998). Highly efficient phosphorescent emission from organic electroluminescent devices, *Nature*, Vol. 395, No 6698, pp. 151-154.

Da Silva Filho, D. A., Kim, E.-G. and Brédas, J.-L., (2005). Transport properties in the rubrene crystal: Electronic coupling and vibrational reorganization energy, *Advanced Materials,* Vol. 17, No 8, pp. 1072-1076.

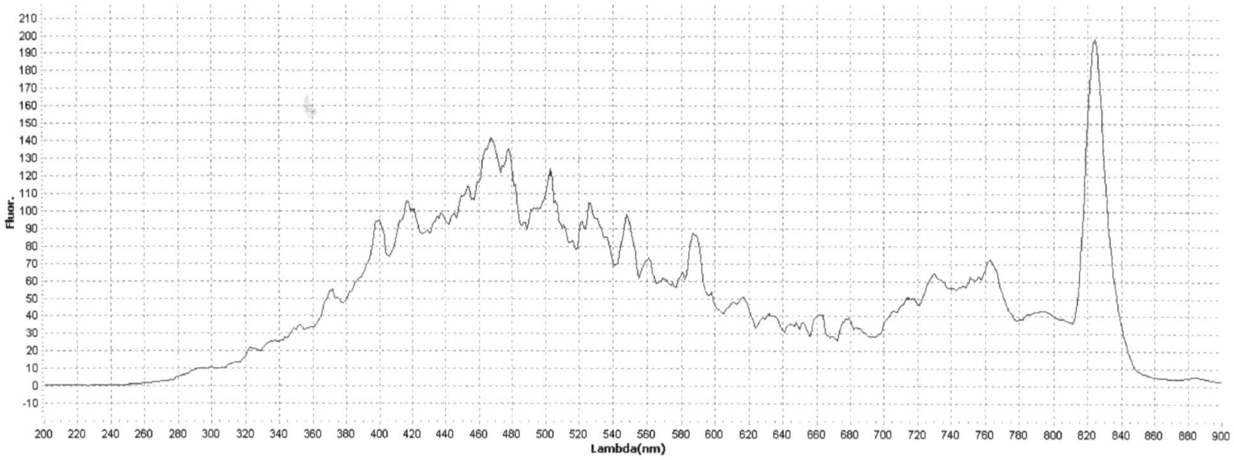

Figure 4. The emission spectrum for the Xe lamp used to excite the studied samples

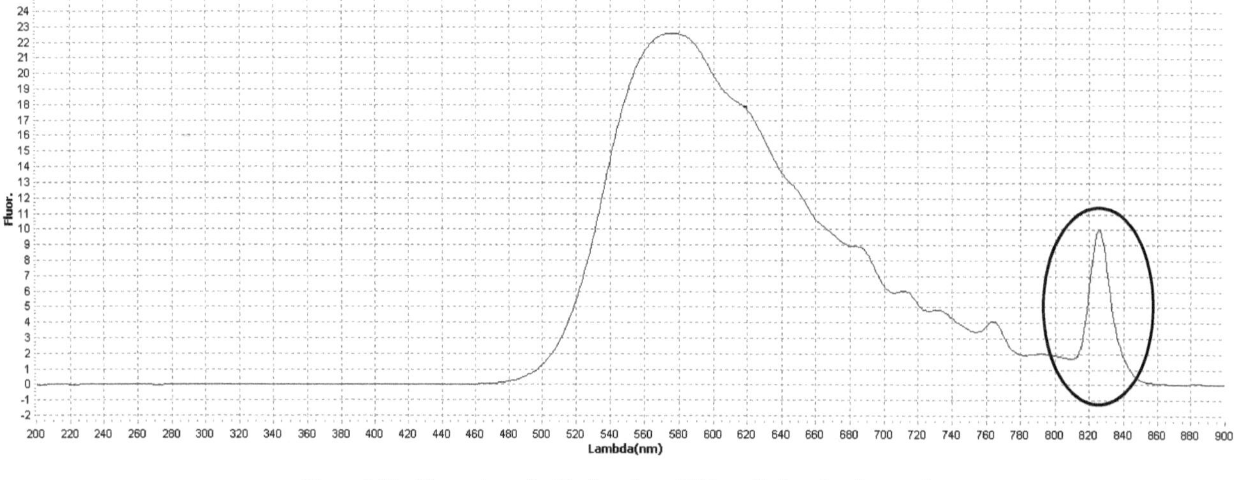

Figure 5. The PL spectrum for (4-ethoxybenyililidene-4′-phenylazobenzene)

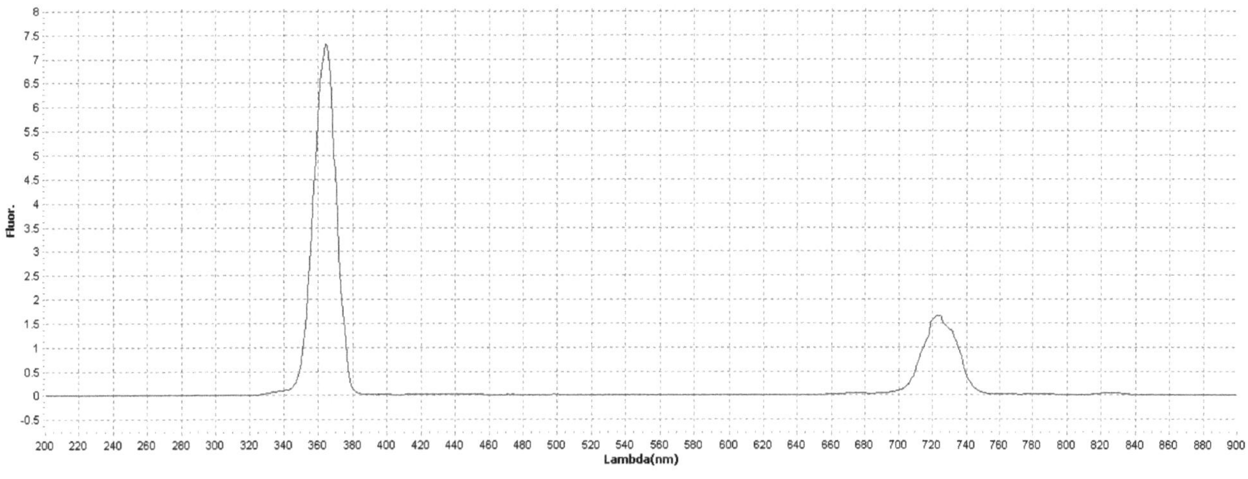

Figure 6. The emission spectrum of filter 365

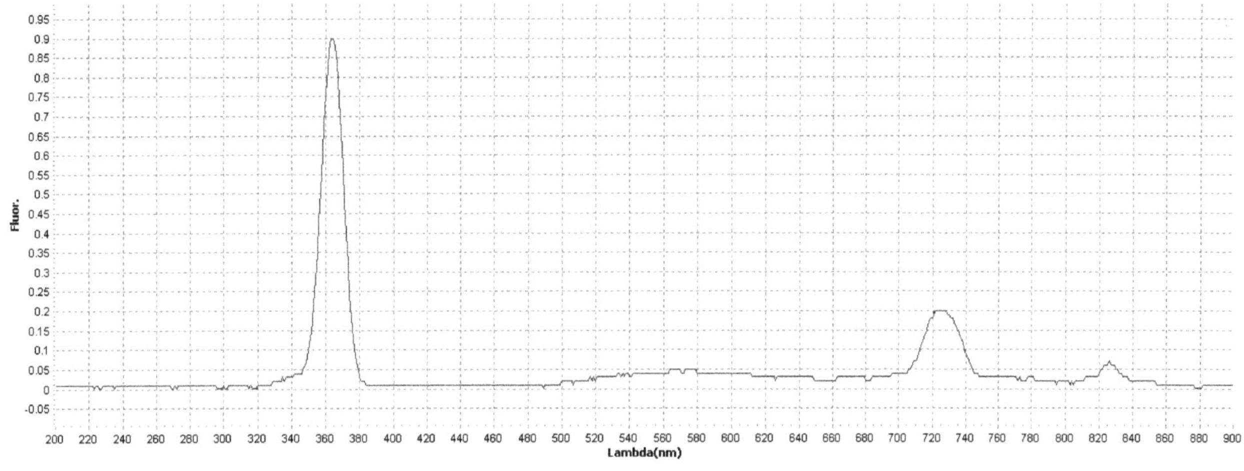

Figure 7. The emission spectrum from sample by using 365 filter during the measurement

Dimitrakopoulos, C.D. and Malenfant, P.R.L., (2002). Organic thin film transistors for large area electronics, *Advanced Materials*, Vol. 14, No 2, pp. 99-117.

Gommans, H., Schols, S., Kadashchuk, A. Heremans, P., and Meskers, S.C.J., (2009). Exciton diffusion length and lifetime in subphthalocyanine films, *Journal of Physical Chemistry C*, Vol. 113, pp. 2974-2979.

Huang, L.S. and Chen, C.H., (2002). Recent progress of molecular organic electroluminescent materials and devices, *Material Science and Engineering Reports*, Vol. 39, pp.143-222.

Lambert, J., Gronert, S., Shurvell, H. and D. Lightner, (2011). *Organic Structural Spectroscopy*, Upper Saddle River: Pearson Education.

Muccini, M., (2006). A bright future for organic field-effect transistors, *Nature Materials*,. Vol. 5, No 8, pp. 605-613.

Petrenko, T. Krylova, O., Neese, F. and Sokolowski, M., (2009). Optical absorption and emission properties of rubrene: Insight from a combined experimental and theoretical study, *New Journal of Physics*, Vol. 11, 23 p., DOI: 10.1088/1367-2630/11/1/015001

Peumans, P., Yakimov, A. and Forrest, S.R. (2003). Small molecular weight organic thin-film photodetectors and solar cells, *Journal of Applied physics*, Vol. 93, No 7, pp. 3693-3723.

Settle, F., (1997). *Handbook of Instrumental Techniques for Analytical Chemistry*, Upper Saddle River: Prentice Hall.

Sharma Y. R., (2012). *Elementary Organic Spectroscopy: Principles and Applications*, Ram Nagar: Chand, S.

Suprvnowicz, Z., Buda, W. Mardarowicz, M. and Patry, A., (1985). Correlations between retention on liquid crystalline stationary phases and chemical structure: I. Dimethylnaphthalenes, *Journal of Chromatography A*, Vol. 333, pp. 11-20.

Tiong, S., Lee, T., Lee, S., Sreehair S. and Win, Y., (2011). Schiff base liquid crystals with terminal iodo group: Synthesis and thermotropic properties, *Scientific Research and Essays*, Vol. 6, No 23, pp. 5025-5023.

Computer Simulation of Glued Connection

Oleg Tatarnikov[1,2]

[1] Rocket and Spacecraft Composite Structures Department, Faculty of Special Machinery, Bauman Moscow State Technical University, 5 2nd Baumanskaya Street, Moscow, 105005, Russia
[2] Higher Mathematics Department, Plechanov University of Economics, 36 Stremyanny Per., Moscow, 117997, Russia

Abstract: The finite element (FE) simulation model using FEMAP Version 10.2.0 software was developed for numerical analysis of a glue connection. The specimen consisting of two ceramic plates connected in an overlapped way was considered as a subject of an analysis. It was shown that the chosen finite element model adequately describes the stress-strain state of the glued joint at large strain. As a result the close fit between experimental data and calculated results was achieved. The results of strain-stress analysis of adhesive connection permitted to obtain the more precise values of mechanical characteristics of the glue material.

Key Words: Glue connection, Adhesive, Finite element model, Large strains, Structural analysis.

1. Introduction

Adhesives are widely used in glued joints of structural elements in aerospace engineering, shipbuilding, and automobile production, as well as for sealing devices in instrument engineering, civil engineering, and many other fields of industry.

There are two main problems related with mechanics of adhesives. The first one is the problem of failure. The second problem is concerned to the strain-stress analysis of glued parts. The questions of adhesive strength have been considered in a lot of articles presented as numerical and experimental result as well. For instance the article (Goncalves, 2003) presents a FE model for three-dimensional progressive failure analysis of adhesive joints. The experimental results are presented in the article (Towse, 1998) where tensile failure strain of a proprietary single-part aerospace epoxy adhesive has been investigated using three different specimen sizes.

The FE model was used to simulate the deformation behavior of thermoplastic olefin (TPO) during tensile test (Ma, 2010). The simulation results gave a clear description of stress distribution and evolution of necking during tensile test.

The general approach to the problems of adhesion and adhesives included mechanics of connection, chemical aspects of glues and surface properties is given by Pocius (2002). The elastic behavior of adhesives the author considered basing on the model of bending of beams (Timoshenko, 2002).

The reliability of glued joints depends on the degree of their numerical and experimental adjustment in the whole range of operational conditions, including large relative deformations that may reach dozens of percents. The quality of calculations is determined by the accuracy of the initial data, most of all by the mechanical characteristics and the accuracy of the numerical model. That is why the cycle of studies has been carried out for the purpose to establish the correspondence between the experimental and calculated data.

2. Shear testing of glue connection

The elastic characteristics of the glue layer were determined in the course of the shear testing of two ceramic plates connected by an overlap. The schematic diagram of testing is shown in Fig. 1. The upper element of joint 1 is subject to force F and lower element 3 is motionless. As a result of this influence, the upper element is shifted by T_x along the X axis, whose direction coincides with that of the vector F. The following experimental dependence was obtained:

$$\boldsymbol{F} = \boldsymbol{F}(\boldsymbol{T_x}) \quad (1)$$

Experiments show that dependence (1) has a pronounced nonlinear character, which is determined mainly by large strain of the glue layer. The typical plot of dependence (1) is shown in Fig. 2. Three regions can be conditionally separated on the diagram as follows: ***I*** is the region of nonlinear-elastic deformation, ***II*** is the region of the beginning of the process of the microdestruction of the glue, and ***III*** is the region of the joint destruction. Considering the problem based on the mechanics of continua, the behavior of the joint corresponding to the first and second regions of the diagram can be simulated.

3. Simulation model

The selection of FE model was based on the obvious assumption that the prevailing strain component in the glue layer is shear strain - γ_{xy} (coordinate system is shown in Fig.1). Thus, the FE model of the glue layer should be three-dimensional, despite the relatively small thickness of the layer as compared to other dimensions of it. Most often, the adhesive thickness in the glued joint is smaller than 0.4 mm. Depending on the type of solved problem, the glued elements can be considered in both two- and three-dimensional formulations. In the case under consideration the stress- strain state of ceramic elements was not a goal of the analysis that is why they were modeling by two-dimensional plate elements.

The adhesive interaction between the glue and the ceramic elements are taken into account by introducing two boundary layers into the simulation model between the ceramic elements and the glue layer (Fig. 3).

It was supposed that two materials namely material of the glue filling roughness onto the element surface and the ceramic material were distributed chaotically in the boundary layer. Thus, the material boundary layer was simulated as the composite material whose Young's modulus E_{bl} is determined according mixture rule

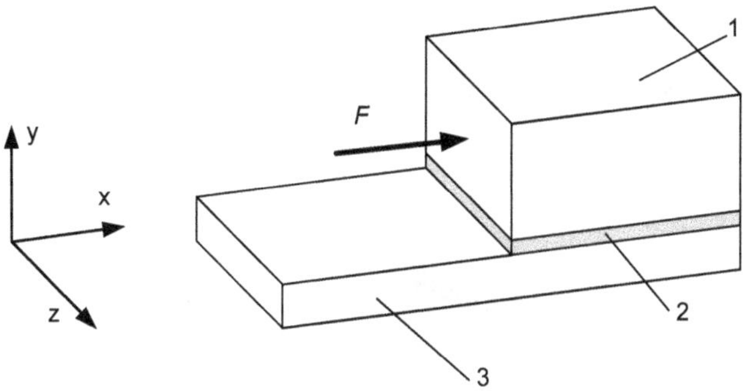

Figure 1. Schematic diagram of glued joint: 1 - top element, 2 - glue layer, and 3 - bottom element of the joint.

$$E_{bl} = E_c\left(1 - C_g\right) + E_g C_g \qquad (2)$$

were E_c - ceramic Young's modulus, E_g - glue Young's modulus, C_g - volume contents of glue into the boundary layer.

The simulation scheme of the joint had the form shown in Fig. 4.

The specific features of mechanical properties of glue material are as follows:

(i) Sealants are weakly compressible materials with Poisson's ratio of $\nu = 0.4 - 0.5$;

(ii) The shear modulus G of these materials varies in the range 0.3 – 2.0 MPa;

(iii) The ultimate tensile strain reaches 160%.

Due to the fact that the relative translation of glued elements reaches 1.5 mm usually (Fig. 2) while the glue thickness is sufficiently small the calculation of the deformed state of the glued joint should be performed taking into account the large strain effect (Novozhilov, 1963), i.e., the geometrically nonlinear approach. Moreover, it is necessary to take into consideration the physical nonlinearity by which the process of damage accumulation in the boundary layer is simulated.

Considering all of that the finite element model of the glued joint was chosen as follows: this model contains

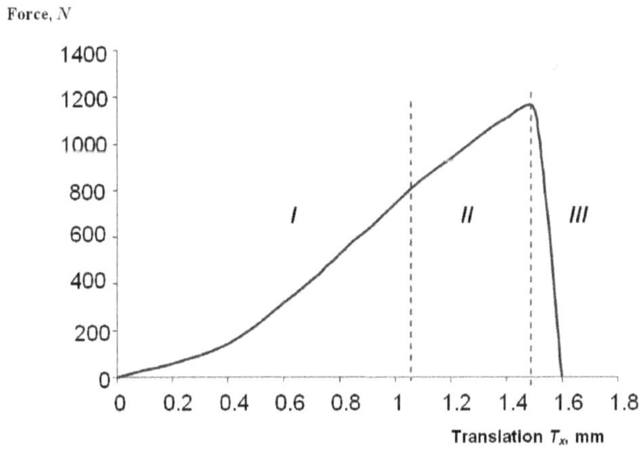

Figure 2. Force–translation diagram.

three types of finite elements: the ceramic material is represented by plate-type elements, the boundary layer is represented by solid-type elements with nonlinear-elastic properties, and the glue layer is represented by solid-type elements with elastic properties (Fig. 5).

The dependence on the numerical results and the mesh size in the perpendicular to the adhesive layer direction was analyzed. The model of a 4×8×0.4 mm glue layer was considered. The following mechanical characteristics were taken:

- for the glue material - shear modulus G = 2 MPa, Poisson's ratio $\nu = 0.48$;
- for the ceramic elements - shear modulus G = 20000 MPa, Poisson's ratio $\nu = 0.3$.

Numerical test showed that the accuracy of calculations in case of translations was not less than 98%. This fact illustrates a plot presented in Fig. 6. In this picture two diagrams "translation – force" corresponding to the mesh sizes $N_h = 2$ and $N_h = 3$ are shown.

3. Numerical results

The results of numerical simulation were compared with experimental data. The dimensions of the glue layer were as follows: thickness × width × length =0.4×20×15 mm.

The testing results are shown in the form of translation – force diagrams in Fig. 7. The sample was 2 times preliminary subject to the load – unload cycles in order to eliminate the influence of technological gaps in the testing fitment.

n Fig. 7 curve "Test no. 1" corresponds to the first loading- unloading cycle, curve "Test no. 2" - to the second one. "Test no. 3" was considered as valid one. The considerable difference of diagrams of the sample testifies the considerable influence of initial gaps in the testing fitment on the testing results. A series of calculations was performed for the glued joint with different values of the shear modulus. The comparison of the obtained numerical values of the force – translation characteristic corresponding to the linear section of the diagram with the data of testing provided the value of the shear modulus equal to G = 0.4 MPa. The value of the Poisson's ratio was determined in a similar way. The results closest to testing were obtained for the Poisson's ratio $\nu = 0.4$. The experimental

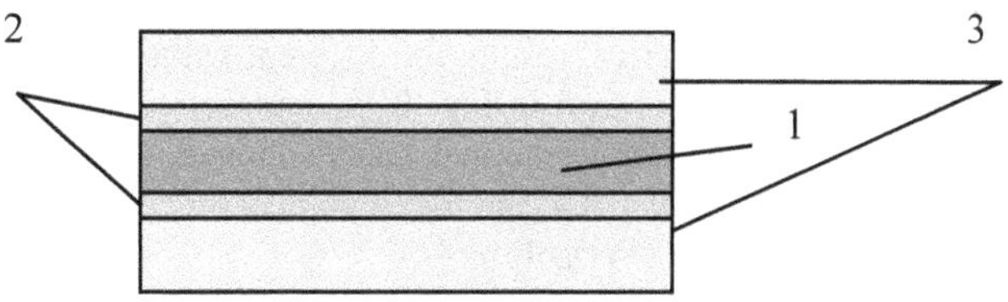

Figure 4. Simulation model of glued joint: 1 - glue layer, 2 - boundary layer, 3 - ceramic.

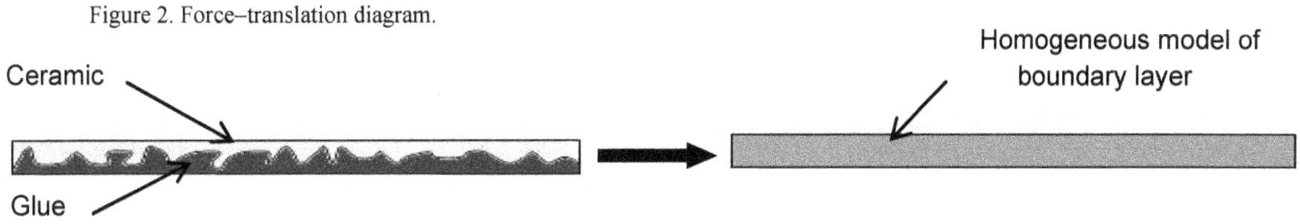

a) b)

Figure 3. Simulation of boundary layer: (a) adhesive–ceramic interface, (b) interface model (boundary layer).

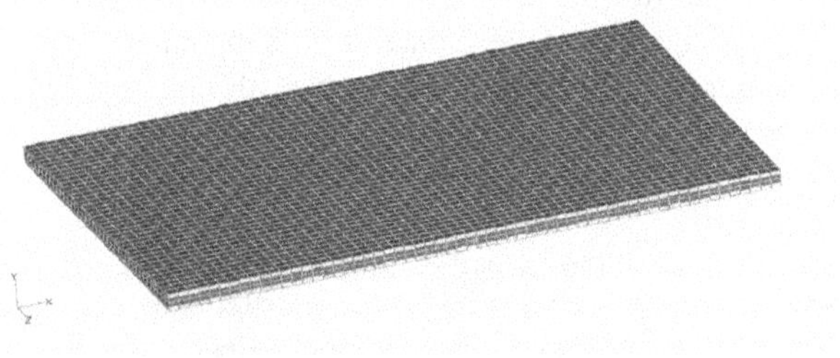

Figure 5. Finite element model of the joint: 1 - ceramic, 2 - boundary layer, 3 - glue layer.

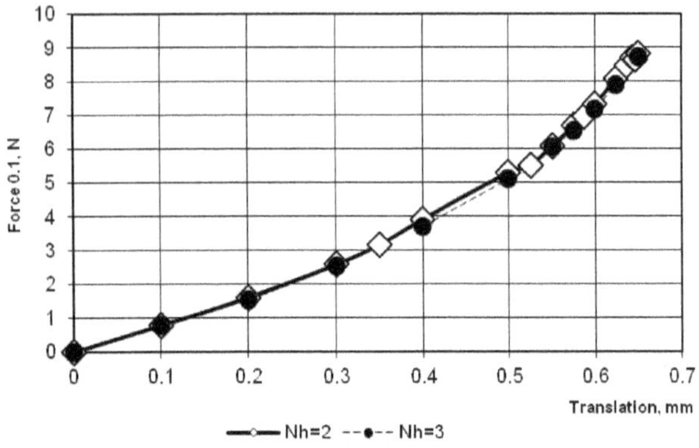

Figure 6. Translation – force diagrams received with mesh size $N_h = 2$ and $N_h = 3$.

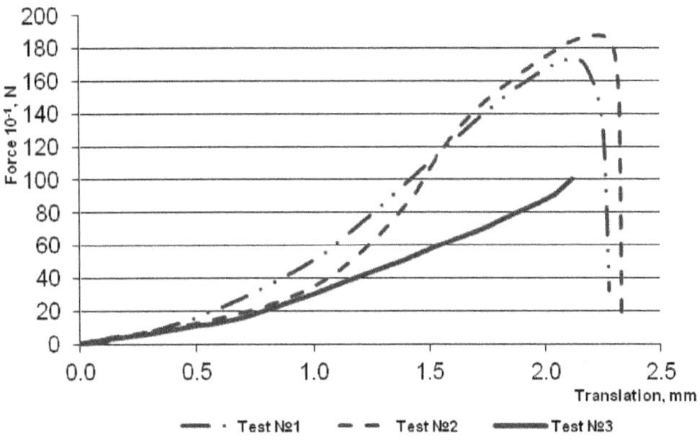

Figure 7. Experimental curves of glued joint testing.

curve and the numerical results for different values of the Poisson's ratio are shown in Fig. 8. A comparison of the experimental data and numerical results showed that, for the Poisson's ratio v = 0.4, the maximal difference for the tensile force was 5.8% (table).

4. Conclusions

Basing on the obtained experimental and numerical data, the following conclusions can be drawn.

(1) The chosen numerical model adequately describes the stress-strain state of the glued joint.

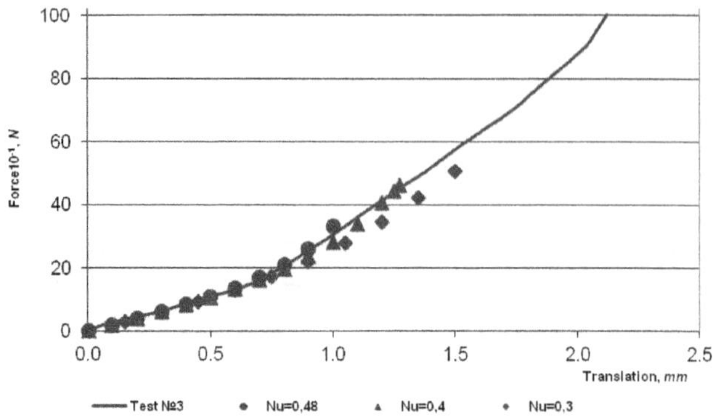

Figure 8. Experimental data and numerical results for the glued joint.

Table 1.
Experimental data and numerical results

Translation, mm	Force, ×10 N			
	Test No.1	Calculation		
		v = 0.48	v = 0.40	v = 0.30
0.004	0.52	0	0	0
0.098	2.62	2	2	-
0.151	3.66	-	-	2.98
0.2	4.57	4.04	4	-
0.301	6.33	6.16	6.12	6.08
0.398	8.7	8.44	8.34	-
0.451	9.93	-	-	9.4
0.499	11.08	10.92	10.72	-
0.601	13.19	13.7	13.34	13.06
0.7	16.08	16.96	16.28	-
0.75	18.23	-	-	17.24
0.801	20.62	20.92	19.66	-
0.899	25.4	26.02	23.6	22.12
1.001	30.47	33.1	28.28	-
1.052	33.17	-	-	27.88
1.101	35.95	-	33.86	-
1.201	41.24	-	40.494	34.6
1.251	43.86	-	44.22	-
1.276	45.21	-	46.18	-
1.351	48.92	-	-	42.3
1.501	57.53	-	-	50.68

(2) The comparison of results obtained using the presented simulation model and experimental data obtained in the course of shear strength testing of the glued joint result in the more precise specification of mechanical characteristics of the glue material.

References

Goncalves, J.P.M., De Moura, M.F.S.F., Magalhaes, A.G. and De Castro, P.M.S.T. (2003). Application of interface finite elements to three-dimensional progressive failure analysis of adhesive joints. *Fatigue and Fracture of Engineering Materials and Structures*, Vol. 26, No 5, pp. 479-486.

Ma, Q., Lai, X., Su, X., Lasecki, J. and Frisch, R. (2010). Modeling and simulation of the large deformation behavior for thermoplastic olefin. *Computational Materials Science*, Vol. 47, No 3, pp. 660-667.

Novozhilov, V. V. (1963). *Foundations of the Nonlinear Theory of Elasticity*. Rochester, New York: Graylock Press.

Pocius, A. V. (2002). *Adhesion and Adhesives Technology: An Introduction*, Munchen: Hanser

Timoshenko, S. P. and Gere, J. M. (2002). *Mechanics of Materials*, Saint Petersburg, Moscow: Lan.

Towse, A., Potter, K., Wisnom, M.R. and Adams, R.D. (1998). Specimen size effects in the tensile failure strain of an epoxy adhesive. *Journal of Materials Science*, Vol. 33, No 17, pp. 4307-4314.

Degradation Behavior and Protection Methods of Polymeric Composite Materials in a Space Environment

Valery Bashkov, Alexey Osipkov, Pavel Mikhalev, Maxim Khafizov

Nanotechnology and Nanomicrosystem Training and Engineering Center, Bauman Moscow State Technical University,
5 2nd Baumanskaya Street, Moscow, 105005, Russia

Abstract: This work provides results of the study of thermal vacuum exposure and exposure of corpuscular stream at epoxy resins with different hardeners as well as a method of protection for spacecraft structural elements made of PCMs, which consists of deposition of protecting diamond-like coatings at their surfaces. There was demonstrated that the carbon coating deposited on the polymeric materials has good barrier properties (practically stopping outgassing under the impact of space factors) and reduces material erosion under exposure of corpuscular emissions.

Key Words: Composites, Diamond-like coating, Outgassing, Space.

1. Introduction

In aircraft building polymeric composite materials (PCMs) were started being used in the 1970s. Replacement of a metal element for analogous one made of polymeric composite enables the reduction of weight by 10 to 50%. At currently manufactured aircraft PCMs take 7 to 25% of weight. And there is a trend to increase usage of PCMs in load-carrying structures of an airframe. As an example the Boeing company can be taken: if in the Boeing 777 aircraft which development was started in 1990s the PCMs are used in about 9% of frame elements, the in the Boeing 787 manufactured in 2010 already 50% of fuselage elements are made of OCMs. The Airbus company uses the technology the same way and declared usage of PCMs is 52% of elements of the components of the fuselage of the Airbus 350 aircraft which is currently under development.

The task to reduce mass-dimensional characteristics of a structure is the sharpest one for the space industry. If for a civil aircraft the cost of weight reduction by 1 kg is about 100$ then for a satellite at a synchronous orbit this cost rises to 10 000$ (Berlin, 1995). However at the present time the usage of structural PCMs in space craft does not exceed 10%. At that the most part of PCMs is used in the inner compartments of the spacecraft and in the instrumental part. For example, in the load-carrying structure of the Space Shuttle PCM were used only for some elements of the cargo compartment doors.

The problem of the limited usage of PCMs in spacecraft structure is that the polymers, in contrast to metals, are more prone to the impact of such space factors as high vacuum, temperature cycling from –70 to +150°C as well as ionizing and corpuscular radiations. These factors initialize the following degradation processes: PCM gas emission due to escape of plasticizers, resin hardeners and other elements, material erosion under radiations. This results in the degradation of the physical-mechanical properties of elements, their embrittlement, loss of geometrical parameters, etc. In addition, the combinations of gas emissions and the destruction of PCMs make a considerable contribution into the internal atmosphere of a spacecraft, generated by products of material outgassing, combustion products of correcting engines, leakages from the inner compartments. Particles forming on the spacecraft's own outer atmosphere impair insulation properties of the vacuum near the spacecraft, provide light scattering as well as settling on different surfaces of the spacecraft impairing the properties of thermoregulating coatings and making optical elements dirty.

Increasing then PCM share in spacecraft requires the study of the impact of different open space factors on the materials of this class and the development of methods and methods of their protection.

This work provides results of the study of thermal vacuum exposure and exposure of corpuscular stream at epoxy resins with different hardeners as well as a method of protection for spacecraft structural elements made of PCMs, which consists of deposition of protecting diamond-like coatings at their surfaces.

2. Test specimens

It should be noted that the open space factors affect mostly the PCM bonding matrix. This is due to fact that initially carbon fiber and glass fiber is less affected by the influence of these factors. Besides, the core material as a rule is covered with a film of bonding matrix, which protects it against direct exposure. Therefore the object of this study is the epoxy bonding matrix which is widely used as a bonding matrix of PCMs. In this study epoxy resin ЭД-20 (foreign analogy is D.E.R. 331 produced by DOW Chemical company) with different hardener types: of cold cure (polyethylenepolyamine – PEPA) and hot cure (diethylenetriamine – DETA).

It should be noted that it is known that hot cure epoxy resins knowingly are less outgassing because the part that has reacted and formed hard chemical linkage substance is greater than in cold cure epoxy resins. In general the PEPA is characterized by still residue of about 60% by weight that in the case of epoxy resin-hardener ratio as 10 to 1 provides about 6% of knowingly unbound particles in the composition, which have to leave the material under thermal vacuum influence.

The test specimens were manufactured in the form of disk with depth 1 to 2 mm and diameter 20 mm in accordance with the method corresponding to the each hardener.

Spectral optical characteristics of metal meshes, such as transmitivity A_v, reflectivity R_v and emissivity ε, with v for frequency, were determined by means of standard optical devices, while integral parameters, such as A/ε, were determined through thermal tests in vacuum chamber with solar simulation (Reznik et al, 2010; Reznik et al., 2011). The temperature dependence for the specific heat capacity of metal mesh was determined with standard devices, such as IT-c-400. The implementation of these methods increased the accuracy of thermal calculations.

3. Test method of thermal vacuum influence

The test method of thermal vacuum influence on the specimens is based on a standard test method for total mass loss and collected volatile condensable materials, which is regulated by ASTM E595 "Standard Test Method for Total Mass Loss and Collected Volatile Condensable Materials from Outgassing in a Vacuum Environment".

The principle of the method consists of the following: the prepared material specimens are put into special containers made of stainless steel. The containers are placed into a cluster box provided with a heating system enabling the maintenance of a constant temperature. The box is placed into a vacuum chamber and after 24-hour annealing the specimens' mass losses were measured.

As the tasks of the study differ from the standard tasks some changes to the method were made.

At first, the specimen annealing temperature is not fixed strictly at 125°C as the standard prescribes. It is necessary to determine dynamic and limits of the outgassing. As the rate of outgassing components depends on a specific material it is necessary, to accelerate the experiment, to increase the annealing temperature up to the maximum possible limit under the condition of staying within a range of self-similarity (maintaining of physical processes within the polymer). In our case the annealing temperature was selected 10...20°C below the maximum allowable operational temperature of the material.

Secondly, the standard method supposes to hold the specimens for 24 hours and then take them out and weigh them.

Taking into account the fact that the equipment used does not allow to monitoring of the dynamic of the outgassing process and determine its limits in the real time, the specimens are periodically taken out, their mass losses are determined and the specimens are put back into the vacuum chamber for the further annealing. As the specimen mass loss is, as a rule, per an exponential law then the hold time is increased for the each new loading. The exact values of the hold time are determined empirically for each material.

As for this study, the total annealing time was about 53 hours. The specimens were taken out at 2, 4, 16, 22, 35 and 41 hour intervals.

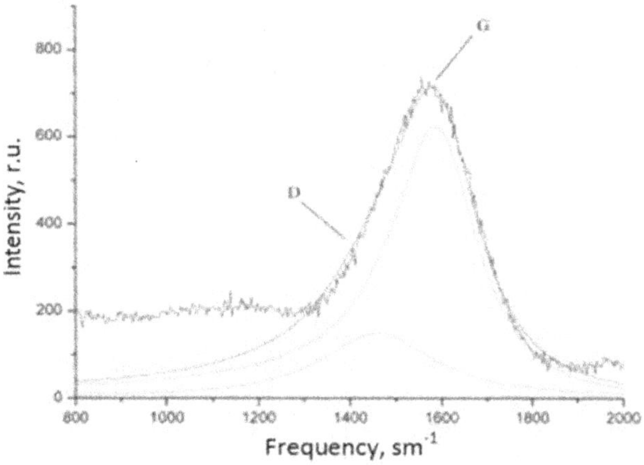

Figure 1. Decomposition of carbon components Raman spectrum of DLC-coating

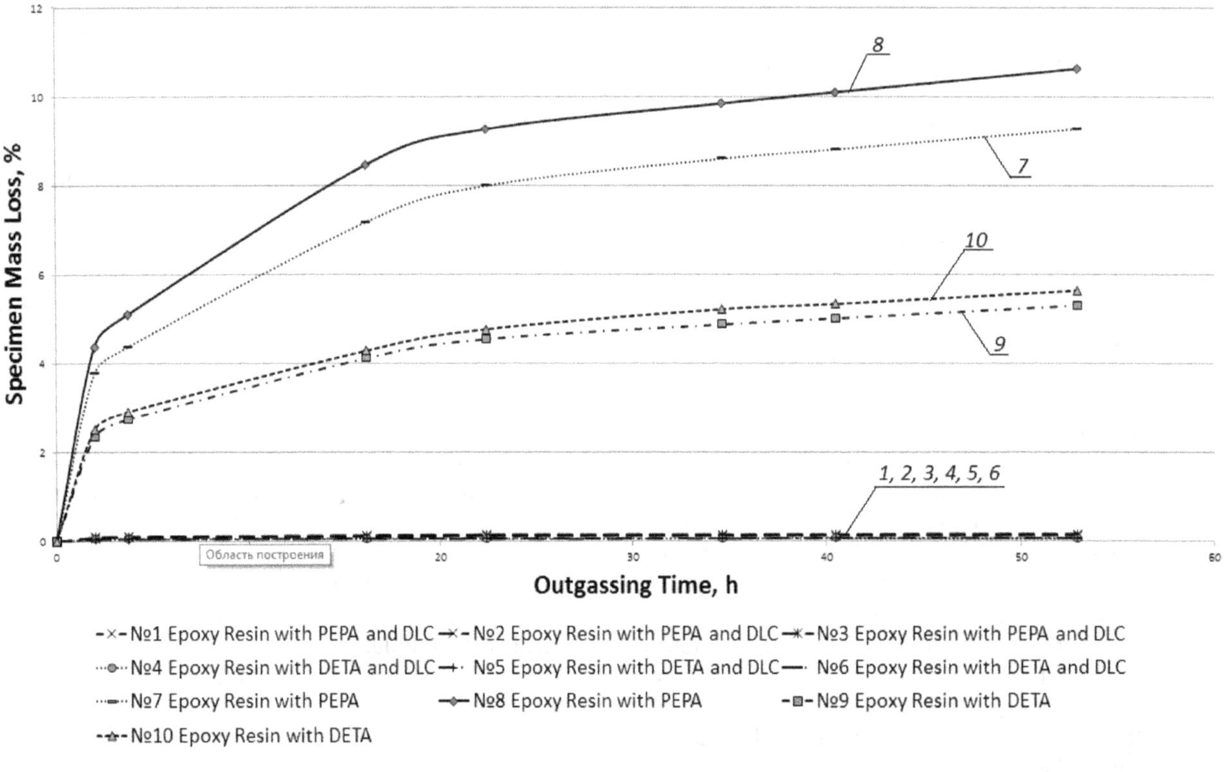

Figure 2. Samples weight loss for 53 hours

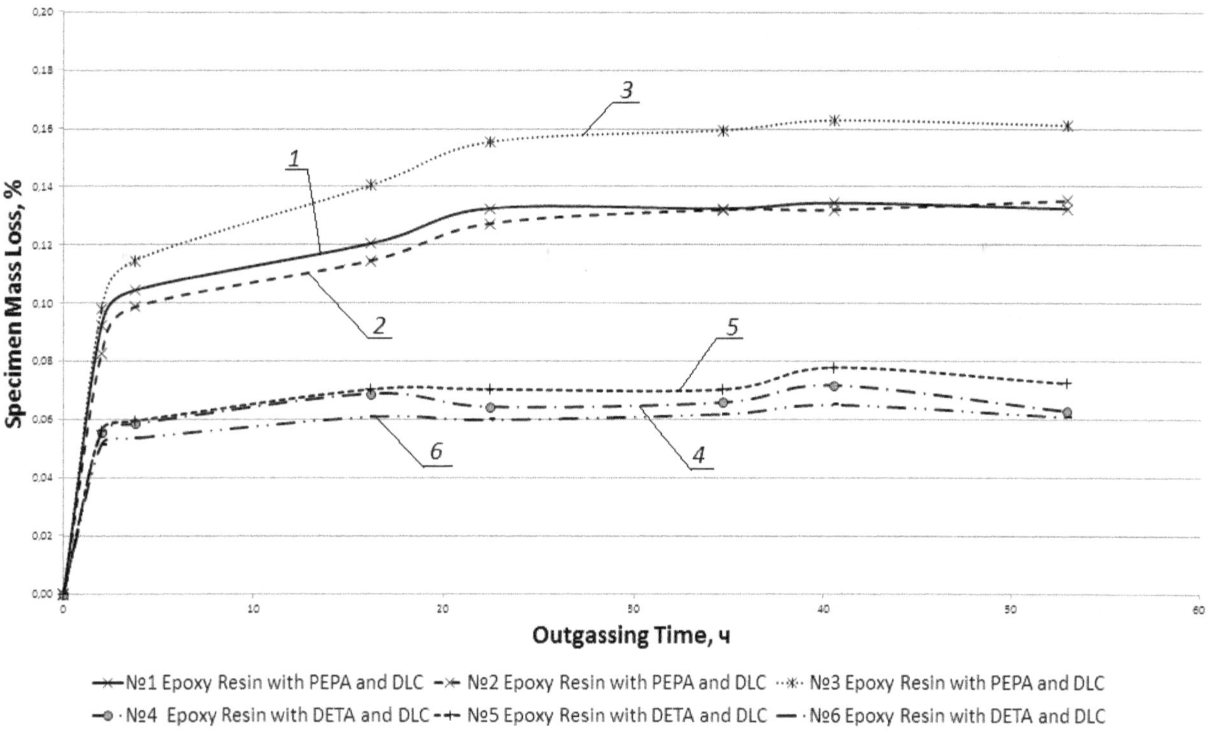

Figure 3. Weight loss of the samples with DLC-coating

In this experimental study of outgassing condensing screens for easy condensable substance were not used although the equipment developed by our team (Yanovich at al., 2012) does enables this process. Research of the condensate formed by outgassing products is an individual task requiring additional researches. It is planned to conduct these researches in the future, including condensing screens not only from stainless steel but from a glass simulating elements of optical systems and specimens with deposited thermoregulating coatings.

Besides research of outgassing processes under thermal vacuum influence there were also conducted research into material specimen resistance to corpuscular radiation. Focused ion beam module "NANOFAB -100" (NT-MDT, Russia) was used as an emission source. The following research method was used: part of the specimen was covered with a screen (foil); the two areas of 250x250 µm with 50% overlapping located at the foil boundary were scanned with the ion beam. It enabled the formation of a distinguished boundary at the material surface and create areas with single and double radiation doses. Then the specimens were studied by the methods of optical and probe microscopy, as well as by the method of scanning electron microscopy (SEM).

4. Deposition of diamond-like carbon coatings

The Diamond-Like Carbon Coatings (DLC) featuring by a number of unique properties such as high mechanical properties, high gas impermeability, biocompatibility, etc. were selected as protecting coatings for PCMs used in spacecraft structure (S. M. Baek at al., 2011). At present polymeric materials coated with layers of DLC coatings are considered as perspective packaging materials in the food industry and in the pharmaceutical industry (Tsubone at al., 2007; Boutroy at al., 2006).

The method of protecting coatings deposition on polymeric materials is based on the developed manufacturing process of multilayer nanostructured diamond-like coatings deposition for micro-size tools (an axial tool with diameter less than 3 mm) at equipment East 01 (New Plasma Technologies LLC, Russia). The instrument surface is coated with matrix multilayer of nanocomposites on the basis of separated electric arc evaporation of carbon from a graphite target with an arc initialized by laser, as well as a separated electric arc evaporation of ions of metals, particularly of titanium and chromium. The deposited coatings protect the material against abrasive agents' action, have high hardness (above 22 GPa) and strength, protect tools against wearing, have low fiction coefficient relative to the tool material that eliminates heating of the side surface, as well as have effect of self-lubrication. During the development of this manufacturing process there were used an approach for the creation of nanocomposite hardening coatings, which is based on the formation of different combination of the following three types of effects: the effect of multiphase hardening, the effects depending on grain size, and the effects suppression of shift of grain boundaries by forming strong contacts.

Development of modes of DLC deposition on polymeric composite materials was based on the points of the experimental design theory with optimization of the manufacturing process by the parameter of number of sp3 -hybridized carbons (f) in the deposited film ($f > 20\%$). At that the metallic phase was dropped due to excessive local heating-up of the polymeric material substrate during heavy metal ion bombardment.

The number of sp3-hybridized carbons was estimated by the method of Raman spectroscopy at equipment Ntegra Spectra (NT-MDT, Russia) provided with a Raman spectrometer.

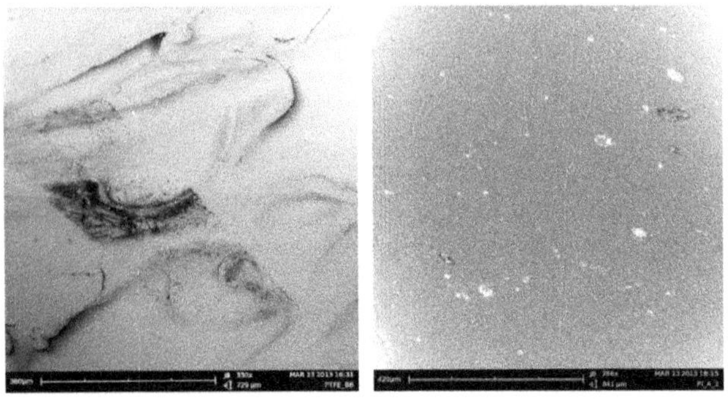

a) b)

Figure 4. Image of the specimen irradiated with ion radiation

The carbon phase of the coating was observed in the range of 1100 to 1700 cm^{-1}. The phase decomposition was conducted per a method described in (Ferrari and Robertson, 2000) (Fig. 1). D-peak was prescribed by Loerntzian, G-peak was prescribed by Breit-Wigner-Fano function. The expansion parameters are the following: peak integral intensity ratio (I_D/I_G) 0.230, percentage of sp^3-hybridized carbons (f) 28%.

The coating provided by the method of plasma electric arc deposition have a high coefficient of similarity to diamond but contain more disordered linkages that is indicated by the more widen D-peak.

Beside evaluation of similarity to diamond the depth and hardness of deposited coatings were also measured at IR-spectral ellipsometer IR-VASE and nanoindentor Hysitron TI750 UBI correspondingly.

In the result of the development the following manufacturing modes were found: laser radiation power – 20 mJ; laser radiation frequency – 20 Hz; cathode voltage* – 250 V; hold-in solenoid current* – 2 A; working gas (Ar) supply* – 15 cm^3/h; coating deposition time* – 60 min; chamber temperature* – 34°C; rate of rotation of specimen table* – 3 rpm; carbon source voltage * – 200 V; chamber pressure – $4 \cdot 10^{-5}$ mbar (* variable parameters).

At these manufacturing modes the depth of deposited coating is 250 nm, hardness is about 17 GPa.

5. Research results

Fig. 2 and 3 show mass loss diagrams for 4 types of specimens: cold cure epoxy resin (hardener PEPA), hot cure epoxy resin (hardener DETA), and the same materials with deposited DLC-coatings with depth 250 nm. As it was expected specimens of cold cure epoxy resin have the largest mass loss (about 10%). As it was mentioned above it is due to the fact that hardener PEPA has still residue as well as to the fact that there are large number of unreacted substance which has no strong chemical bonds with the resin particles. At that it should be noted a wide spread of measurements between two specimens prepared at the same time and under the same conditions. Probably, this is due to the fact that the resin and hardener were not mixed properly. Specimen 8 has less hardener and has resulted in the fact that resin molecules were not bonded fully and they get out during the vacuum annealing process. Specimen 7 has more hardener, naturally there were more reacted resin molecules and in the result the mass loss was less.

For the hot cure specimens 9 and 10 the mass loss was half as high that is explained by absence of still residual in hardener DETA as well as more complete polymerization due to higher temperatures. A wide spread of measurements between these two specimens is considerably less and it can be explained by the difference in specimen geometry and different surface areas. In future it is planned to conduct additional researches of material manufacturing parameters on the larger number of specimens of the same type.

Mass loss of the specimens with DLC-coatings is considerably less than when compared with the specimens of initial materials. At that, the mass loss of hot cure specimens (4, 5 and 6) after the first annealing was 0.06% and later it changed within the range of the statistical uncertainty. Such behavior can be explained by the fact that atmospheric adsorbate went away of the specimen surface and practically full absence of material-own outgassing. The mass losses of cold cure specimens were some higher at level of 0.13–0.16%. At that level results for Specimens 1 and 2 were practically the same and for Specimen 3 it was higher by 0.03%. The latter can be explained by presence of some contamination at the surface.

There by it can be stated that deposition of DLC-coating is extremely effective preventing and releasing of the unbound molecules from the epoxy resin. That is if material selection is regulated by Standard ASTM E595 then the pure epoxy resin cannot be used as PCM matrix neither with hot cure nor with hot cure all the more. At that deposition of a protective coating outgassing is reduced up to the guaranteed level less than 1% even under the more sever conditions of tests (more long-lasted holding, higher temperature).

Ion-beam resistance was studied at epoxy resin specimens of hot cure. One initial material specimen and one specimen with deposited DLC-coating were exposed. Each specimen had two areas to be scanned. The further studies were made by the method of scanning electron microscopy (SEM). Figure 4(a) shows an image of the exposed area of the surface of the epoxy resin without coating, made with the help of SEM FeiPhenom. The erosion area is well viewed, at that it has distinguished boundaries along the protective mask and along other sides of the scanned area. It is explained by well by using a focused ion beam and allows the process to be done without protective masks.

Figure 4(b) shows an image of the surface area of a specimen with DLC-coating supposedly at the area of ion beam exposure. The fact is that in spite of the markers it is not possible to identify the exposed place because there are no any traces of the marker. In future it is planned to enhance the marker system to enable the location of the precise position in the area exposed by ions. It will enable further study of the exposed area not only by the SEM method but the probe method also and that will give more data on specimen resistance to ion exposure.

5. Conclusion

There was proposed a method of polymeric materials protection against open space factors. There was developed a method of deposition of Diamond-Like Carbon coatings to surfaces of polymeric materials by means of a vacuum electric arc process with plasma beam separation in order to protect the polymeric materials against open space factors. For the test specimens there was demonstrated the following:

- the carbon coating deposited to the polymeric materials has relatively high coefficient of similarity to diamond (coefficient of similarity to diamond is more than 0.2), high hardness (about 20 GPa);
- the coating has good barrier properties (practically full stop of outgassing);
- the coating reduces material erosion under exposure of corpuscular emissions.

Acknowledgments

The work was done as a part of the contract № 11.519.11.3035, 12 March 2012.

References

Baek, S.-M., Shirafuji, T., Saito, N. and Takai, O. (2011). Fabrication of transparent protective diamond-like carbon films on polymer, *Japanese Journal of Applied Physics*, Vol. 50, 5 pp., DOI: 10.1143/JJAP.50.08JD08

Berlin A. A. (1995). Modern polymeric materials (PCMs), *Soros' Educational Journal*, No 1.

Boutroy, N., Pernel, Y., Rius, J. M., Auger, F., von Bardeleben, H.J., Cantin, J.L., Abel, F., Zeinert, A., Casiraghi, C., Ferrari, A.C. and Robertson, J. (2006). Hydrogenated amorphous carbon film coating of PET bottles for gas diffusion barriers, *Diamond and Related Materials*, Vol. 15, pp. 921-927.

Ferrari, A.C. and Robertson, J. (2000). Interpretation of Raman spectra of disordered and amorphous carbon, *Physical Review B*, Vol. 61, No 20, pp. 14095-14107.

Tsubone, D., Hasebe, T., Kamijo, A. and Hotta A. (2007). Gas barrier properties and periodically fractured surface of thin DLC films coated on flexible polymer substrates, *Surface and Coatings Technology*, Vol. 201, No 14, pp. 6423-6430.

Yanovich, S.V., Litvak, Y.N., Mikhalev, P.A. and Bashkov, V.M. (2012). Development of machine for the study of resistance of materials to the effects of space factors, *Engineering Bulletin*, Vol. 11. [Online]. Avalable at http://engbul.bmstu.ru/doc/511815.html

Pedagogical and Linguistic Innovations in an ESP Course in Composite Materials and Technologies

Inna Shafikova

English for Machine Engineering Department, Faculty of Linguistics, , Bauman Moscow State Technical University,
5 2nd Baumanskaya Street, Moscow, 105005, Russia

Abstract: The article is concerned with the issue of English for Specific Purposes (ESP) courses for university students. The evolution of language teaching methodology in general, the development of ESP and the evolution of teaching languages at universities in Russia are compared and contrasted. The emphasis is laid on the current trends in the education process in Russia, in particular the Bologna convention aimed at unifying and standardizing systems across Europe. The English language course in composite materials at Bauman MSTU is reviewed and analyzed, the requirements for an improved ESP course are presented, and the content of the prospective course is outlined.

Key Words: Composite materials and technology, Course design, English for Specific Purposes, Methodology

1. Introduction

People have been learning and teaching foreign languages ever since they became aware of the other nations' existence. Being proficient in one or several foreign languages has always been considered an essential part of good education. Teaching and learning foreign languages for the greater part of its history as we know it has employed the Grammar-Translation Method, which in turn had its origins in the teaching of Latin. It was assumed that a language could be mastered by learning its vocabulary and a set of grammar rules. The limitations of this approach–lack of aural fluency, absence of the links between the language taught and the real-life situations and low learners' motivation – called for new approaches to language instruction. One of them, the Direct Method, involved learning the language in the natural, inductive way. It did not become widely spread in the state education institutions due to the absence of set methodology and procedures. The Communicative Method was introduced in the 1970-s and gained ground quickly. Its main aim is to develop the communicative competence through meaningful learning activities. The course is based on functions or notions rather than on grammatical structures.

The communicative approach still remains the most common method used in the modern classroom, though it is supplemented with selected features of the other approaches. One of the main characteristics of the modern classroom is greater tolerance of errors and mistakes and understanding of their inevitability and necessity. In addition 'there is considerably less emphasis on translation, there have been moves towards increased use of the target language as the medium of instruction and towards broadening the range of learning activities to include oral presentations, group discussions, debates, précis, summaries, letters, reviews and reports" (Gray and Klapper, 2003).

2. Characteristics of English for Specific Purposes worldwide and in Russia

he phenomenon of teaching a language for specific purposes (LSP) is a fairly recent one, which can be accounted for by the growing need for international professional communication. This paper is focused on teaching English for Specific Purposes (ESP) as opposed to teaching General English (GE), because it is the English language that has become the predominant medium of cross-cultural interactions – 'the key to the international currencies of technology and commerce' (Hutchinson, 1987).

The methodology of ESP became the centre of attention in the 1970-s. There appeared a generation of learners who clearly knew what they wanted, which shifted the focus from the teacher to the learner and led to the emergence of Learner-Centred Approach. 'Time and money constraints created a need for cost-effective courses with clearly defined goals' (Hutchinson, 1987). English for Science and Technology (EST) was the first variant of English to be described and analyzed and it has set the trends for the development of ESP; other directions of ESP are Business English and English for Social Sciences.

It is widely and erroneously believed among the teachers of English as a Second Language that ESP is simply a set of lexical items and a set of grammatical structures typical of this field – so to develop an ESP course in composite materials and technologies, for example, would require selecting the vocabulary of various manufacturing processes, a number of verbs, names of different pieces of equipment and suchlike and supplement it all with the structures of the "technical English", that is passive voice and complex noun phrases. However, this approach would not lead to an ESP course. What distinguishes an ESP from a non-ESP course is the notion of **purpose** and it was emphasized by the leading theorists in the field of English Language Teaching. Dudley-Evans and St. John (1998) based on the work of Strevens (1988) suggested three absolute and five variable characteristics of ESP (Table.1).

The methodology of teaching English at university level in Russia has undergone several stages which coincide with the general trends in the second language teaching as described above. The first course books of foreign languages for different industries were developed in the 1950s-60s accompanied by the development of the theory of teaching foreign languages at non-linguistic universities. Reading was considered to be the key skill, probably due to the fact that no other media were available at that time, and a specialist would hardly be expected to need or to be able to communicate in the target language. A foreign language was viewed as a cognitive, not a communicative tool. Consequently, the course books concentrated mainly on teaching different types of reading, on mastering the vocabulary of the relevant field and understanding various syntactical structures of the target language (Pavlova, 2011). The main medium of instruction was the first language, i.e. Russian, and translation was both the main learning technique and the main objective of teaching.

Things have changed somewhat in the 1980s, when other media of communication became available and there emerged the need for full-scale interactions in the target language. The course books started to include listening comprehension, speaking and writing communicative activities.

3. Current trends in the Russian education system

The present day state of affairs calls for reevaluation of the existing methods and courses within the paradigm of the contemporary education. The most important factor is the integration of Russia into the global education system – the Bologna process. This initiative is directed at unifying the European education to ensure its quality and uniformity across nations as a response to globalization. Russia became part of the process in 2003 by signing the convention on the mutual recognition of diplomas and certificates and is now striving to achieve the objectives of the Bologna convention: mobility of students and teachers, compatibility and comparability of diplomas and qualifications, growth of autonomy and competitiveness of European higher education institutions due to the integration of intellectual potential. The Bologna doctrine also provides for the principles and methods to meet the objectives specified: system of credits, adequate accreditation procedures, adoption of a two-tier architecture of academic degrees, transparency of information flows for all education modules, and implementation of the new diploma supplements.

Table 1
Absolute and variable characteristics of an ESP course

Strevens	Dudley-Evans and St. John
Absolute characteristics	
meeting specified needs of a learner	meeting specific needs of the learner;
related in content to particular disciplines, occupations and activities	
	using the underlying methodology and activities of the discipline it serves;
centred on language appropriate to those activities in syntax, lexis, discourse, semantics, etc.	focusing on the language (grammar, lexis, register), skills, discourse and genres appropriate to these activities.
Variable Characteristics	
restricted as to the learning skills to be learned	
	related to or designed for specific disciplines
may not be taught according to any pre-ordained methodology	a different methodology from that of general English (not necessarily)
	designed for adult learners, either at a tertiary level institution or in a professional work situation
	intermediate or advanced students
	basic knowledge of the language system

The implications of the Bologna process for Russia are manifold. First of all, the recognition of Russian degrees worldwide would mean better access to employment on the global market and wider opportunities for academic exchange. Secondly, it would increase Russia's compatibility in the international education market. It would also allow Russia to reform its education systems by aligning its academic standards with the European level and preserving the achievements of the past (Pimenova, 2006).

The implications of the Bologna process on the English language teaching are equally significant. Academic mobility would require higher proficiency in one or several European languages, English being the obligatory one. This would result in changes in the English language courses at University level. The general trends are as follows:

- connection with the professional activity, i.e. conducting the needs analysis to create a customized ESP course;
- life-long learning, i.e. designing a range of courses to enable continuous improvement of the professional communicative competence;
- diversified language learning system, i.e. providing for groups of students with different aims and different abilities;
- democratization of higher education, i.e. providing equal access to the foreign language courses to the students with varying starting abilities;
- fundamentalization of higher education, i.e. combining a General English course and an ESP course for various levels of study;
- integration, i.e. unification and standardization of the foreign language courses and increased academic mobility;
- humanization, i.e. utilizing an axiological approach to meet the personal and professional demands of the learner;
- information technology, i.e. using IT and multimedia to promote autonomous learning;
- learner-centered approach; i.e. accounting for the learners' various needs and planning the course accordingly (Polyakova, 2011).

4. Course of English for Specific Purposes at Bauman MSTU Composites department. Current state and perspectives

Bauman Moscow State Technical University aims to be at the forefront of the academia and industry. Composite materials are one of the perspective directions of development: the Engineering Education and Research Centre for Composite Materials has been opened recently and there is a project underway to create a world-level Laboratory and Centre of Expertise in Composite Materials with the collaboration of the leading researchers and industry specialists, academic exchange is gaining momentum, and plans for dual degrees have been introduced, namely, dual PhD and MEng programmes with Glyndwr university, UK. All these trends make the students and teachers all the more aware of the importance of the foreign language competence in the relevant professional field. The English language curriculum is being currently revised and adapted to the recently introduced modular system. The undergraduate course of English lasts seven terms and with two class hours per week it totals 3 modules or 34 class hours a term. There are 3 modules in the term, each module focusing on one general engineering field of interest and one grammar problem.

The methodological rationale is that the English language course complements the course of studies at the university: the first four terms are concerned with the general engineering subjects, and the English language course suggests a broad range of texts on different topics (automobiles, aircraft, underwater craft, lasers, electronics etc). Grammar is presented in isolated chunks and the main objective of the grammar syllabus is the ability to correctly identify and translate the structures. The selection of grammar taught is determined firstly, by the need to review the grammar learned at school – such broad concepts as the Simple, the Continuous and the Perfect tenses. Both very narrow grammar aspects, such as participle, or gerund, or infinitive, and very broad aspects, such as conditionals, are all allocated the same amount of class hours. Reading is the dominating skill to use and to practice, in spite of the texts having relatively low informative value – with some information outdated (especially the texts on electronics, automobiles, IT). Comprehension is achieved and evaluated by means of translation (Orlovskaya et al, 2007).

The merits of this course are its correspondence to the Russian learners' needs, a solid foundation in grammar, focus on morphology, and a wide range of topics covered. The emphasis on translation can be justified by the specifics of the grammar translation method, although it requires a careful reappraisal in the view of various methods of computer-assisted translation systems. A very serious shortcoming of this course is the absence of writing and listening activities, which the linguistics department attempts to compensate for by introducing students' reports and presentations into the course to increase the students' motivation and provide them with a certain extent of autonomy in selecting the areas and topics for reading and discussion. Experience shows that most students respond well to such type activities.

Cambridge English for Engineering course is taken in the fifth and the sixth terms. This course is a direct opposite to what is taught in the preceding four semesters. First of all, it utilizes a functional syllabus practicing the communicative skills that future engineers might need in their career, e.g. describing component shapes and features, explaining tests and experiments, discussing performance and suitability, etc. Secondly, there is no apparent emphasis on reading – the texts are quite short and serve primarily as a source of information. Thirdly, the course offers a wealth of listening and speaking activities based on everyday engineering situations (Ibbotson). The universality and general engineering character of this course can be considered a disadvantage as there is no connection with the specific engineering field. Another drawback is a lack of grammar practice, which remains an important issue for students at this level. This course is supplemented by independent background reading, when students autonomously select, read and translate texts in their discipline. A tentative approach is being taken to teach students essential academic writing skills by means of producing summaries and précis of their background reading texts.

The seventh term gives the students the opportunity to encounter and use the language in the context of their specific engineering field. By that time the students have supposedly gained greater professional competence and built a solid foundation in General and General Engineering English. There exist courses developed by the linguistics department of the Bauman MSTU for every major. The composite materials department students are offered *Guidelines in Teaching Reading Specialized Literature on*

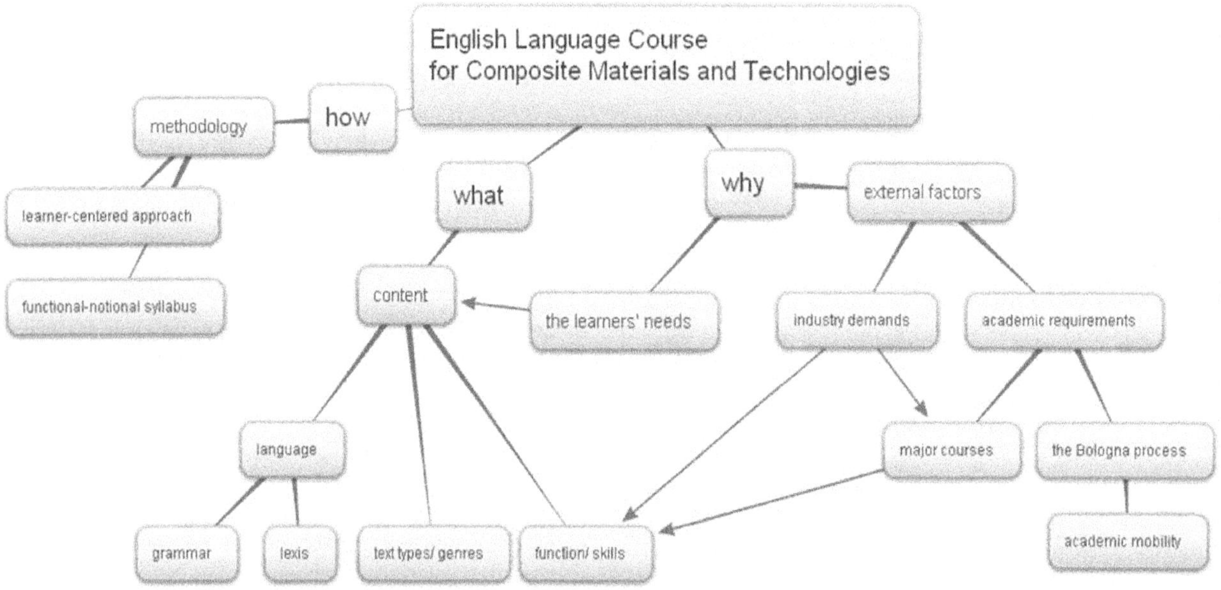

Figure 1. The ESP course for Composite Materials and Technologies' Content and Rationale.

Table 2
Proposed content of the ESP course in Composite materials and technologies

Undergraduate major	Function	Lexis	Grammar
Technology of processing and modifying new materials	Explaining needs, problems and solutions; stating objectives;	Composites manufacturing processes verbs	Modals, Infinitives of purposes
Fundamentals of physics, chemistry, and fabrication of composites	Describing a process	Physics, chemistry terms	Conditionals
CAD basics	Explaining dimensions; working with drawings	Specifications of dimensions	
History of research in composites	Sequence of events		Past tenses
Composite structures mechanics	Explaining forces	Physical forces	
Principles of scientific research	Describing methods, giving definitions		Relative clauses
Polymer composite structures development	Describing use or function; describing shape and appearance	Composite materials, properties. Shapes of composite parts	Noun attributes
Technology of reusable space vehicles	Describing parallel processes	Stages in a composites fabrication process	Passive voice
Composite media mechanics	Describing motion, forces	Verbs to denote movement	
Optimization of composite structures and procedures	Expressing purpose; expressing strengths/ weaknesses	Processes sequence/ simultaneity; Types of composites fabrication equipment	Purpose clause

Composite Materials. The methodology and structure of these courses are in line with the grammar translation method, which means the comprehension is achieved via translation into the first language.

Summing up all of the above, the following conclusions can be drawn as regards the English language course for composite materials and technologies (Fig.1), which must:

- correspond to the students and the prospective employers' needs.
- complement the specialized engineering disciplines;
- reflect the latest developments in the composite materials industry;
- utilize communicative approach and be based on a functional-notional syllabus;
- train students to receive professionally oriented instruction in English as part of the academic mobility programme.

The content of the ESP course for composites should be examined in greater detail. Industry demands and the major disciplines are to determine the choice of functional units to be included in the course. Below is the list of undergraduate specialized courses and proposed functional, lexical, and grammatical items (Table 2).

Table 2 is an outline of the proposed course for the Composites department. A detailed needs analysis is to be conducted to determine the exact content and the structure of the course. The key stakeholders are not only the linguists and English language teachers, but also the Composites department academics and the Composite Centre executives. The relevant communicative and professional skills are to be thoroughly analyzed and selected. They may include presentation skills, specific writing skills (technical reports, grant proposals, e-mails), academic cognitive skills (note taking), certain skills to assist in job hunting and career building (interviews, writing CVs, cover letters).

Conclusion

It is certain that the reality of the global world make the second language competence a requirement rather than an option. A good engineer must be an integral part of the international professional community, and this cannot be achieved without the shared communication medium – which happens to be the English language. A number of current trends in foreign languages learning affect course design and course content selection. In particular, life-long and content-integrated learning place the emphasis on accuracy and not on fluency. Emergence and continuous improvement in computer-assisted translating systems mean that conventional approaches to teaching reading should be reviewed and reinvented. The Bologna process aimed at unifying education systems across Europe promotes academic mobility, which means that the second language turns from a cognitive to a communicative tool. All of the above calls for the development of a new ESP course corresponding to the relevant needs of today's learners and utilizing the advantages of the cutting-edge technology and methodology. A course of English for Specific Purposes at universities should first of all meet demands of the students, academia and industry and should be designed by means of all the stakeholders' collaboration.

References

Dudley-Evans, T.(1998). *Developments in English for Specific Purposes: A Multi-Disciplinary Approach.* Cambridge University Press.

Gray, C. and Klapper, J. (2003) Key Aspects in Teaching and Learning in Languages.*A Handbook for Teaching & Learning in Higher Education*, London and New York: Rouledge Falmer

Hutchinson, T. and Waters, A. (1987). *English for Specific Purposes: A Learner-Centered Approach.* Cambridge University Press.

Howatt, A.P.R. and Widdowson, H.G. (2004). *A History of English Language Teaching.* Cambridge University Press.

Ibbotson, M. (2008). *Cambridge English for Engineering.*Cambridge University Press.

Ivanova, L.I. (2001). *Guidelines on Teaching Reading Technical Literature on Manufacturing Technology of Composite Parts*. Moscow, Bauman MSTU.
Joint Declaration on Harmonization of the Architecture of the European Higher Education System. [online]. Accessed http://www.bologna-berlin2033.de/en/main_documents/index.htm
Kuzmina, L.A. and Shashmurina A.V. (2012). *Teaching to Read Specialized Literature on Composite Materials in English*. Moscow, Bauman MSTU.
McDonough, J. (1984). *ESP in Perspective: A Practical Guide*. London: Collins ELT.
Orlovskaya, I.V., Samsonova L.S. and Skubrieva A.I. (2002). *The English Language Course Book for Engineering Universities*. Moscow, Bauman MSTU.
Pavlova, I.P. (2011). Modern foreign language manual for teaching non-linguists: Problems and perspectives. *Vestnik of Moscow State Linguistic University*, No 12, Vol. 618, Priorities in Foreign Language Teaching, pp. 43-60. (in Russian)
Pimenova, N.Yu. (2006). *On Strategy of Promotion of Russian Education on the International Market, Education in Russia for Foreigners*.WorldWideWeb. [online]. http://en.russia.edu.ru/information/ananlyt/947/ Accessed 26/03/2013..
Polyakova, T.Y. (2011). Trends of modernizing the foreign language training system in engineering education. *Vestnik of Moscow State Linguistic University*, No 12, Vol. 618, Priorities in Foreign Language Teaching, pp. 9-18. (in Russian)
Strevens, P. (1988). ESP after twenty years: A re-appraisal. In M. Tickoo (Ed.), *ESP: State of the Art*. Singapore: SEAMEO Regional Centre.

www.ingramcontent.com/pod-product-compliance
Lightning Source LLC
Chambersburg PA
CBHW082302210326
41520CB00049B/7129